Cambridge IGCSE® & O Level

Essential
Chemistry
Workbook

Third Edition

Roger Norris

W0036216

OXFORD
UNIVERSITY PRESS

Great Clarendon Street, Oxford, OX2 6DP, United Kingdom

Oxford University Press is a department of the University of Oxford. It furthers the University's objective of excellence in research, scholarship, and education by publishing worldwide. Oxford is a registered trade mark of Oxford University Press in the UK and in certain other countries

© Oxford University Press 2021

The moral rights of the authors have been asserted

First published in 2021

All rights reserved. No part of this publication may be reproduced, stored in a retrieval system, or transmitted, in any form or by any means, without the prior permission in writing of Oxford University Press, or as expressly permitted by law, by licence or under terms agreed with the appropriate reprographics rights organization. Enquiries concerning reproduction outside the scope of the above should be sent to the Rights Department, Oxford University Press, at the address above.

You must not circulate this work in any other form and you must impose this same condition on any acquirer

British Library Cataloguing in Publication Data
Data available

978-1-38-200619-4

3 5 7 9 10 8 6 4

Paper used in the production of this book is a natural, recyclable product made from wood grown in sustainable forests. The manufacturing process conforms to the environmental regulations of the country of origin.

Printed in India by Gopsons Papers Ltd., Sivakasi

Acknowledgements

®IGCSE is the registered trademark of Cambridge International Examinations.

The publishers would like to thank the following for permissions to use their photographs:

Cover image: Jirik V/Shutterstock

Artwork by Aptara, QBS Learning & OUP

Although we have made every effort to trace and contact all copyright holders before publication this has not been possible in all cases. If notified, the publisher will rectify any errors or omissions at the earliest opportunity.

Links to third party websites are provided by Oxford in good faith and for information only. Oxford disclaims any responsibility for the materials contained in any third party website referenced in this work.

This workbook is designed to accompany the *Essential Chemistry for IGCSE* student book. It is designed to help you develop the skills you need in order to help you do well in your IGCSE Chemistry examination. The book follows the order of the chapters in *Essential Chemistry for IGCSE*. Each page of questions provides additional questions related to each double page in the student book.

The questions focus on the areas you need to know about for your exam:

- Knowledge (memory work) and understanding (applying your knowledge to answer questions about familiar or unfamiliar situations or substances).
- Handling information from data, tables, and graphs.
- Solving problems (including chemical equations and chemical calculations).
- Experimental skills and investigations.

The first 20 units include a range of question types that you will come across in your chemistry examinations:

- Choosing words to complete sentences: you are usually given a list of words to choose from. This will help you learn and remember key facts.
- Putting statements in the correct order or selecting the correct statement from a list.
- Testing your ability to understand chemical formulae and to construct equations.
- Undertaking chemical calculations involving reacting masses, concentration, empirical and molecular formulae, and percentage yield.
- Some questions ask you to interpret data from diagrams, graphs, and tables. Others ask you to interpret the results of investigations that may be unfamiliar.
- Some pages include questions involving extended answers. These will help you organise your arguments and understand the depth of answer that is needed.

Other important features of this workbook that should help you succeed in chemistry include:

- An introductory Language Lab section in each of the first 20 units, which focuses on scientific words. These are often placed in a particular context. Examples include fill-in-the-gap exercises, word searches, and crosswords.
- A unit focusing on language and the importance of identifying key words in questions. This includes vocabulary practice as well as practice in reading and analysing questions.
- A unit focusing on how to make the most of revision time through active revision and mind mapping.
- A unit on mathematics for chemistry. This includes practice in writing formulae, rearranging expressions, working through calculations, and drawing graphs.
- A unit on practical aspects of chemistry including planning an experiment (the selection of apparatus and materials and working safely), measuring, recording data, and drawing graphs. It also includes analysis of results and evaluation. This is followed by a unit suggesting how these aspects of practical chemistry can be applied to projects.
- A selection of IGCSE-style questions of the type that are set in the theory papers will help you to see connections between different parts of the syllabus.
- Full answers are given to all the questions.
- A glossary to help you understand the meaning of important chemical terms.

We hope that the range of differing exercises in this workbook will help you develop your skills in and understanding of chemistry and help you succeed in this subject.

Contents

Contents

Language lab

Complete the following sentences about the particles in solids, liquids and gases using words from the list.

close	everywhere	far	fixed	move	regularly	sliding	vibrate

In solids the particles are .... arranged and close to each other. The particles only

.... They do not .... from place to place. In liquids, the particles are not

arranged in a .... pattern and are .... together. The particles move by

.... over each other. In gases, the particles are .... apart and are able to move

.... rapidly.   [8]

1. Box **A** shows the arrangement of 7 particles in a gas. Complete the boxes **B** and **C** to show the arrangement of 16 particles in a solid and 16 particles in a liquid.

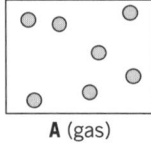  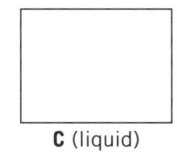

      A (gas)   **B** (solid)   **C** (liquid)   [4]

2. Statements **A, B, C** and **D** are about the properties of solids, liquids or gases. Link each statement to the correct state by writing either solid, liquid or gas after each statement.

    A It takes the shape of its container and has a surface ....

    B It spreads everywhere throughout the container ....

    C It has a definite shape ....

    D It can be poured from one beaker to another beaker ....   [4]

3. On the axes below draw graphs to show (**a**) how the volume of a gas changes when the pressure is gradually increased and (**b**) how the volume of a gas changes as the temperature is increased.

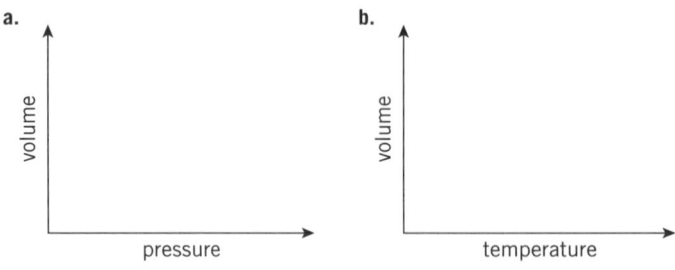

                                                                                                                        [3]

4. Complete these sentences about the kinetic particle theory using words from the list.

attract	directions	spheres

Particles in gases behave as hard .... They move in all .... They do not

.... each other.   [3]

Language lab

Complete the following sentences about melting and boiling using words from the list.

| attraction | boiling | energy | escape | melting | surface | vibrate | weak |

When a solid is heated, the increase in .... makes the particles .... more.

The forces of .... between the particles are weakened. At the .... point,

these forces are .... enough for the particles to be able to move and slide over each

other. When the liquid is at its .... point, the particles have enough energy to

.... from the .... of the liquid.   [8]

1. Complete the diagram by writing the names of the changes of state **A**, **B**, **C**, and **D**.

[4]

2. Box **Y** shows particles of gas in a container with a plunger.

   a. Draw a diagram in box **Z** to show what happens when the gas is compressed.   [1]

   b. Use the kinetic particle theory to explain why the pressure in **Y** is less than the pressure in **Z**.

   ....

   ....

   .... [3]

3. 10 cm^3 of a gas is placed in a gas syringe. Explain, using the kinetic particle theory the effect of increasing the temperature of this gas. The atmospheric pressure is constant.

   ....

   ....

   .... [3]

Language lab

Complete the following sentences about a cooling curve using words from the list.

constant   decreases   freezes   kinetic   released   temperature

When a liquid above room temperature cools the .... energy of the particles ....

The .... of the liquid falls. At the melting point the temperature stays .... for a

time. This is because thermal energy (heat) is being .... when a liquid .... [6]

1. The table shows the melting points and boiling points of three substances.

substance	melting point / °C	boiling point / °C
ethanol	−117	79
methane	−182	−164
naphthalene	81	218

a. Which substance has the lowest melting point? .... [1]

b. Which substance is a solid at room temperature? Explain your answer.

.... [2]

c. Which substance is a liquid at room temperature? Explain your answer.

....

.... [2]

2. The diagram shows a heating curve for substance T.

What is the physical state or states of T at the following points?

A .... [1]

B .... [1]

C .... [1]

D .... [1]

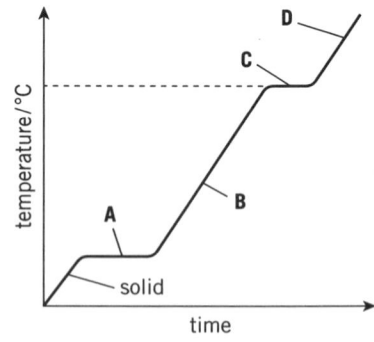

3. When a gas cools to form a liquid the temperature falls at first and then remains constant for a short time before decreasing again. Explain why the temperature falls then remains constant.

....

....

.... [4]

Match the words below to the phrases **A** to **E**.

aqueous   saturated   solute   solution   solvent

A   A substance which dissolves a solute. ....   [1]

B   A substance which dissolves in another substance. ....   [1]

C   A mixture of a solute and a solvent. ....   [1]

D   Dissolved in water. ....   [1]

E   Contains as much as possible of something. ....   [1]

1. State the colour changes when:

   a. Water is added to anhydrous copper(II) sulfate.

      from .... to ....   [2]

   b. Water is added to anhydrous cobalt(II) chloride.

      from .... to ....   [2]

2. State the meaning of these terms:

   a. hydrated .... [1]

   b. water of crystallisation .... [1]

3. Change these volumes into dm^3.

   a. $250 \, cm^3$ .... [1]   b. $150 \, cm^3$ .... [1]

   c. $50 \, cm^3$ .... [1]   d. $25 \, cm^3$ .... [1]

4. Calculate the concentrations of these solutions in g / dm^3.

   a. 5 g of sodium hydroxide in $500 \, cm^3$ solution. .... [1]

   b. 15 g of sodium chloride in $400 \, cm^3$ solution. .... [1]

   c. 2 g of potassium sulfate in $40 \, cm^3$ solution. .... [1]

   d. 0.5 g of potassium chloride in $25 \, cm^3$ solution. .... [1]

5. Complete this sentence:

   Concentration of solution can be measured in g / dm^3 or .... $/ dm^3$.   [1]

Complete the following sentences about diffusion using words from the list.

changing   gases   hit   liquids   mixed   movement   particles   randomly

In liquids and gases, the .... are constantly moving and .... direction

when they .... other particles. We say that they move .... Diffusion is

the random .... of particles in any direction so that they get .... up.

Diffusion in .... is faster than in .... because the particles move faster

in gases.   [8]

1. A student placed a crystal of a blue dye at the bottom of a beaker of water. After 5 minutes, the crystal disappeared. After 1 day, the solution was blue throughout.

   a. Name the process taking place when the colour spreads throughout the water.

   .... [1]

   b. Use ideas about moving particles to explain why the colour spreads throughout the water.

   ....

   ....

   .... [3]

2. A diffusion experiment is set up as shown.

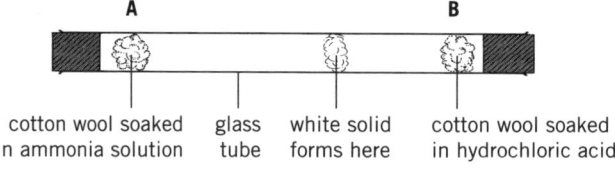

   cotton wool soaked   glass   white solid   cotton wool soaked
   in ammonia solution   tube   forms here   in hydrochloric acid

   Ammonia solution forms ammonia gas. Hydrochloric acid forms hydrogen chloride gas. Explain why the white solid forms and why it is closer to **B** than **A**.

   ....

   ....

   .... [3]

3. The relative molecular masses of three molecules are given.

   ammonia = 17   chlorine = 71   ethane = 28

   Which one of these diffuses most slowly in air? Explain your answer.

   .... [2]

Language lab

Search for eight pieces of laboratory apparatus in this word square. Words may go upwards, downwards, backwards, or forwards, but not diagonally. Put your answers in the space below.

E	S	R	E	K	A	E	B
L	Y	R	R	A	L	T	A
A	R	O	P	N	O	T	L
M	I	F	I	D	N	E	A
O	N	L	P	E	T	R	N
S	G	A	E	R	R	U	C
T	E	S	T	T	U	B	E
I	N	K	T	E	R	M	H
A	S	R	E	M	I	T	E

[8]

1. The diagram shows four pieces of laboratory glassware.

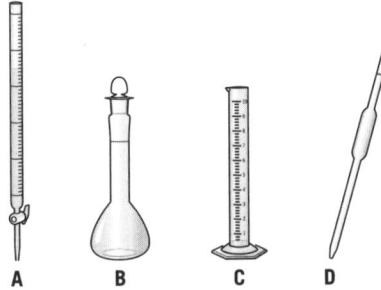

A   B   C   D

a. Name each of these pieces of glassware

A ....   B ....

C ....   D ....   [4]

b. Which piece of glassware would you use to

i. Make up a solution of sodium hydroxide accurately, ... [1]

ii. Deliver 25.0 cm³ of hydrochloric acid? .... [1]

2. The diagram on the right shows part of a burette. Where should you position your eye (direction **A**, **B**, **C**, or **D**) to get a precise reading? Ring the correct answer.

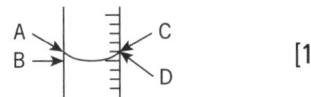

[1]

Language lab

Complete the following sentences about chromatography using words from the list.

attraction   filter   locating   mixture   separate   solubilities   spraying

The method of separating a .... of coloured substances using .... paper

is called chromatography. The colours .... if they have different .... in

the solvent and different degrees of .... for the filter paper. Chromatography can

also be used to separate colourless substances. These are shown up after chromatography

by .... the paper with a .... agent.   [7]

1. Paper chromatography can be used to separate a mixture of dyes.
    Complete the diagram on the right to show the apparatus set up
    for chromatography. Show the position of the baseline.
    Label your diagram.

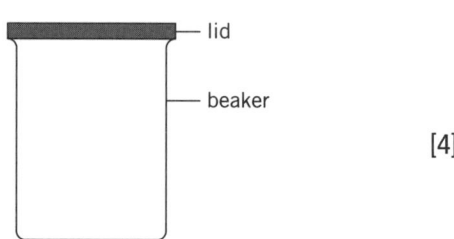

    [4]

2. A paper chromatogram of some amino acids from a mixture of amino acid is shown.
    Two pure amino acids, Ser and Gly were also run on the same piece of paper.

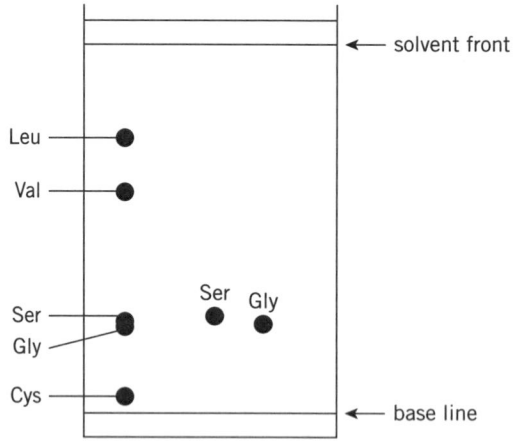

    a. Why was the base line drawn in pencil and not in ink?

    .... [1]

    b. How many amino acids have been completely separated? .... [1]

    c. Which amino acids have not been separated? .... [1]

    d. Calculate the R_f value of Val. .... [1]

    e. Lysine has an R_f value of 0.14. On the diagram above, draw the approximate position of Lys. Label it Lys.   [1]

Complete the following sentences about pure and impure substances using words from the list.

boiling   **decreased**   **exact**   **impure**   **increased**   **pure**   **range**

The melting and .... points of .... substances are sharp. They melt and boil at

.... temperatures. The melting and boiling points of .... substances are not

sharp. They melt over a .... of temperatures. The boiling point of a liquid is

.... if impurities are present. The melting point of a liquid is .... if

impurities are present.   [7]

1. **a.** Which two of these substances are most likely to be pure?
   Underline the correct answers.

   air   aspirin tablets   orange juice

   oxygen gas   sodium chloride crystals   tap water   [2]

   b. Why is distilled water and *not* tap water used in chemistry experiments?

   .... [1]

   c. Seawater is a mixture.

   Suggest a value for the melting point of seawater. .... [1]

2. **a.** Sulfur melts at 119 °C and boils at 445 °C.

   Draw lines between the boxes on the left and the boxes on the right to complete the sentences.

	..... melts over a 4 °C temperature range
Pure sulfur	..... turns to a vapour at 450 °C
Impure sulfur	..... solidifies at 119 °C
	..... has a sharp boiling point [2]

   b. Solder is a mixture of tin and lead which is used to join metals.
   The melting point of tin is 232 °C. The melting point of lead is 328 °C.
   Solder melts at 183 °C.

   i. Why does solder have a lower melting point than either tin or lead?

   .... [2]

   ii. Suggest an advantage of the low melting point of solder.

   .... [1]

Language lab

Link the words **A** to **E** on the left with the descriptions **1** to **5** on the right.

| A Decanting | 1 A substance which dissolves in a solvent. |

| B Filtrate | 2 A mixture formed from a solute and solvent. |

| C Residue | 3 Pouring off a liquid from a solid which has settled. |

| D Solute | 4 In filtration, the liquid which goes through the filter paper. |

| E Solution | 5 In filtration, the solid which remains on the filter paper. |

[3]

1. a. Complete the diagram by writing the correct labels on the dotted lines.

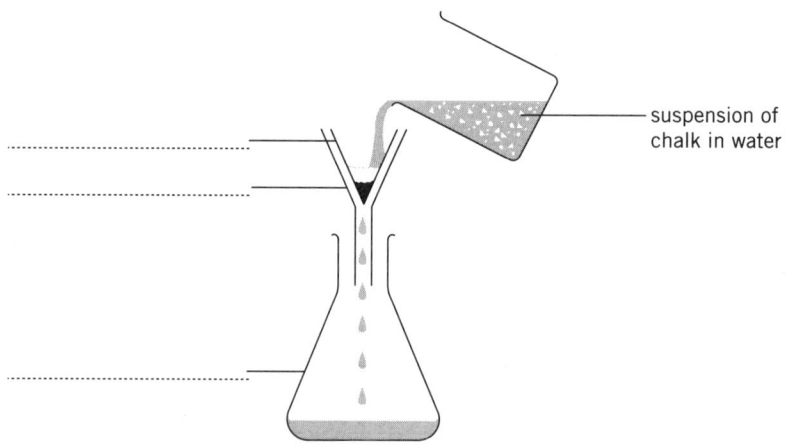

suspension of chalk in water

[3]

   b. On the diagram above label the residue and the filtrate.   [2]

2. a. Put these statements about the crystallisation of zinc sulfate in the correct order.

   A Filter off the crystals
   B Heat the solution to concentrate it
   C Leave the crystals to dry.
   D Wash the crystals with a small amount of organic solvent
   E Leave the solution to cool and form crystals
   F By seeing if crystals form on a cold surface
   G Check that a saturated solution has formed

   Order .... [2]

   b. Suggest why should you should **not** use distilled water to wash the crystals.

   .... [1]

Complete the following sentences about fractional distillation of alcohols using words from the list.

boiling	condenser	further	higher	liquid
lower	receiver	temperatures	vaporised	volatile

There is a range of .... in the distillation column, ... at the

top and .... at the bottom. When .... the more ....

alcohols move .... up the column than the less volatile alcohols. When

the alcohol reaches the .... it changes from vapour to ....

The alcohols are collected one by one in the .... , those with the lower

.... points condensing before those with higher ones.   [10]

1. a. i. Label the diagram of the distillation apparatus to show: **A** the distillation flask; **B** the distillate; **C** the condenser; and **D** where cold water enters.

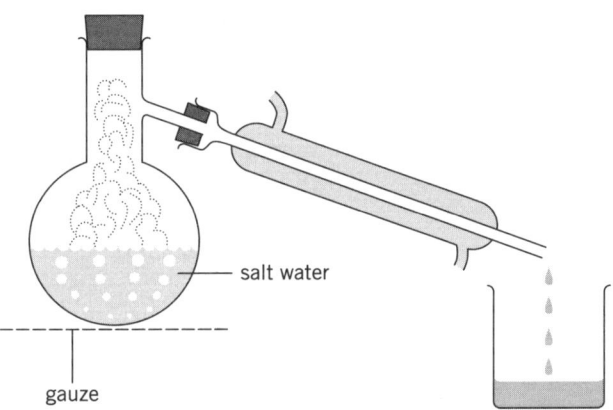

salt water

gauze   [4]

ii. On the diagram above, draw an arrow to show where heat is applied.   [1]

b. i. Explain why this method can be used to separate salt from salty water.

....   [1]

ii. Explain why this method cannot be easily used to separate two liquids which have similar boiling points.

....   [1]

2. Name the separation method you could use to separate:

a. Sand from a mixture of water and sand, ...   [1]

b. Water from a solution of copper sulfate in water, ...   [1]

c. Two liquids with different boiling points. ....   [1]

Language lab

Complete these sentences about atoms using words from the list.

arranged   chemical   electrons   levels   neutrons   nucleus   shells   smallest

Atoms are the .... particles of matter which can take part in a

.... change. Each atom consists of a .... made up of

protons and .... Outside the nucleus are the .... These are

.... in electron .... or energy ....   [8]

1. In 1906, J.J. Thomson suggested a model of the atom shown below.

   How does this model of the atom differ from
   the simple model of the atom we use today?

   sphere of positive charge ——— electrons

   ....

   ....

   .... [3]

2. **a.** Give the sign of the charge (if any) on:

   i. an electron ....   **ii.** a neutron ....   **iii.** a proton ....   [3]

   b. A proton has a relative mass of 1.

   i. Suggest the meaning of the term relative mass.

   .... [1]

   ii. Give the relative mass of one electron. .... [1]

3. **a.** Define proton number .... [1]

   b. Give another name for proton number .... [1]

   c. State the relationship between proton number and the arrangement of the elements in the Periodic Table.

   .... [1]

Language lab

Complete the definition of isotopes using words from the list.

atoms   compound   electrons   element   mass   molecules   neutrons   protons

Isotopes are .... of the same .... with the same number of

.... but different numbers of ....   [4]

1. Three isotopes of hydrogen are:

   $_1^1H$ $_1^2H$ $_1^3H$

   a. Deduce the proton number of hydrogen. .... [1]

   b. What is unusual about the isotope $_1^1H$ compared with other isotopes? ....

   .... [1]

2. Deduce the number of protons, neutrons and electrons in each of these atoms or ions.

atom or ion	number of protons	number of neutrons	number of electrons
$_{17}^{35}Cl$			
$_{58}^{136}Ce$			
$_{11}^{23}Na^+$			
$_{15}^{31}P^{3-}$			

[12]

3. In the space on the right give the symbol for an isotope of lanthanum, La, which has 57 protons and 82 neutrons.   [1]

4. Deduce the relative atomic mass of a sample of rhenium, Re, which contains two isotopes.

   $_{75}^{185}Re$ (abundance = 37.1%) and $_{75}^{187}Re$ (abundance = 62.9%)

   Give your answer to four significant figures.

   [3]

Complete these sentences about electrons using words from the list.

configuration   electron   group   one   outer   period   seven   two

The arrangement of the electrons in shells is called the electron ....

An atom of fluorine has nine electrons, ... in the first shell and .... in

the second shell. Atoms of elements in the same .... have the same

number of electrons in their .... shell. As we move across a ...., each atom

has .... more .... in its outer shell than the element before it.   [8]

1. Complete the table to show the electron configuration of the atoms shown.

element	number of electrons in an atom	electron configuration
nitrogen		
oxygen		
fluorine		
neon		
sodium		
argon		
calcium		

[8]

2. Draw the electron configuration of these atoms. Show all the electron shells. Draw the electrons in pairs where possible.

aluminium	carbon	chlorine	helium
magnesium	neon	phosphorus	potassium

[8]

Link each of these phrases to make a sentence describing an element and a sentence describing a compound.

..... types of atom.....	..... one type of atom.....
..... is a substance containing only.....	..... which are chemically.....
..... combined (bonded).	..... which cannot be broken down further.....
..... is a substance containing two or more.....	..... by chemical means.

An element ....

....

.... [2]

A compound ....

....

.... [2]

1. Complete the table to show the difference between a compound and a mixture using words from the list. Some words may be used more than once.

   any   average   combined   definite   different   elements   physical   present   separated

compound	mixture
The .... cannot be .... by .... means.	The substances in it can be .... by .... means.
The properties are .... from those of the .... which went to make it.	The properties are the .... of the substances in it.
The elements are .... in a .... proportion by mass.	The substances can be .... in .... proportion by mass.

[6]

2. The diagram shows six different substances. Each circle represents an atom. Classify these as elements, compounds or mixtures.

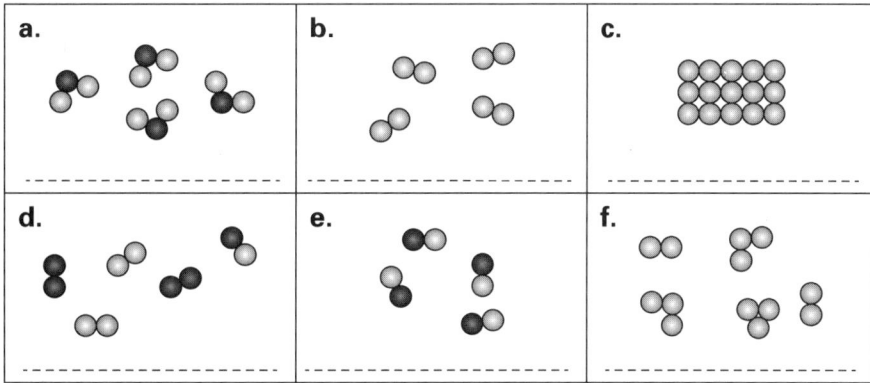

[6]

Search for six words which describe the physical properties of iron. Words may go upwards, downwards, backwards, or forwards, but not diagonally. Put your answers in the space below.

[6]

A	L	I	S	E	D	G	R	A	M
L	M	A	L	L	E	A	B	L	E
I	C	E	G	I	N	L	P	U	R
K	I	N	R	T	S	I	I	S	O
D	O	D	A	C	E	E	L	T	L
C	O	N	D	U	C	T	S	R	B
A	S	E	S	D	U	R	E	E	A
X	I	S	U	O	R	O	N	O	S

1. a. The list below gives some properties of metals and non-metals. Underline the properties which are characteristic of nearly **ALL** metals.

   brittle   conducts electricity   ductile   dull   high melting point

   high density   insulator   malleable   shiny   strong   [4]

   b. The table give some properties of diamond (carbon) and sodium.

diamond (carbon)	sodium
melts above 3550 °C	melts at 98 °C
poor conductor of electricity	good conductor of electricity
good conductor of heat	good conductor of heat
shatters when hit	malleable

   i. Give one way in which diamond behaves as a typical non-metal.

   .... [1]

   ii. Give one way in which diamond behaves differently from a typical non-metal.

   .... [1]

   iii. Give one way in which sodium behaves as a typical metal.

   .... [1]

   iv. Give one way in which sodium behaves differently from a typical metal.

   .... [1]

2. The table shows some properties of three metals.

metal	density in g/cm³	melting point / °C	electrical conductivity in $\Omega^{-1}m^{-1}$	relative strength
aluminium	2.70	660	0.41	7
copper	8.92	1038	0.54	13
iron	7.86	1535	0.11	21

   a. Which metal is best for making the body of an aircraft? Explain your answer.

   .... [2]

   b. Which metal is best for making electrical wiring? Explain your answer.

   .... [2]

Complete the passage about ionic structures using words from the list.

alternate   forces   giant   lattice   negative   positive   strong

A sodium chloride .... is a regular arrangement of .... sodium

ions and .... chloride ions which .... with each other. The ions

are held together by .... ionic attractive .... This structure is called a

.... ionic structure.   [7]

1. Put a ring around the electron configurations which are stable atoms.

   2,8   2,5   2,8,8   2,8,8,2   2   2,8,3   2,8,18,8   [2]

2. Complete the diagrams below to show the electronic arrangement of the stable ions.
   Include brackets and charges.

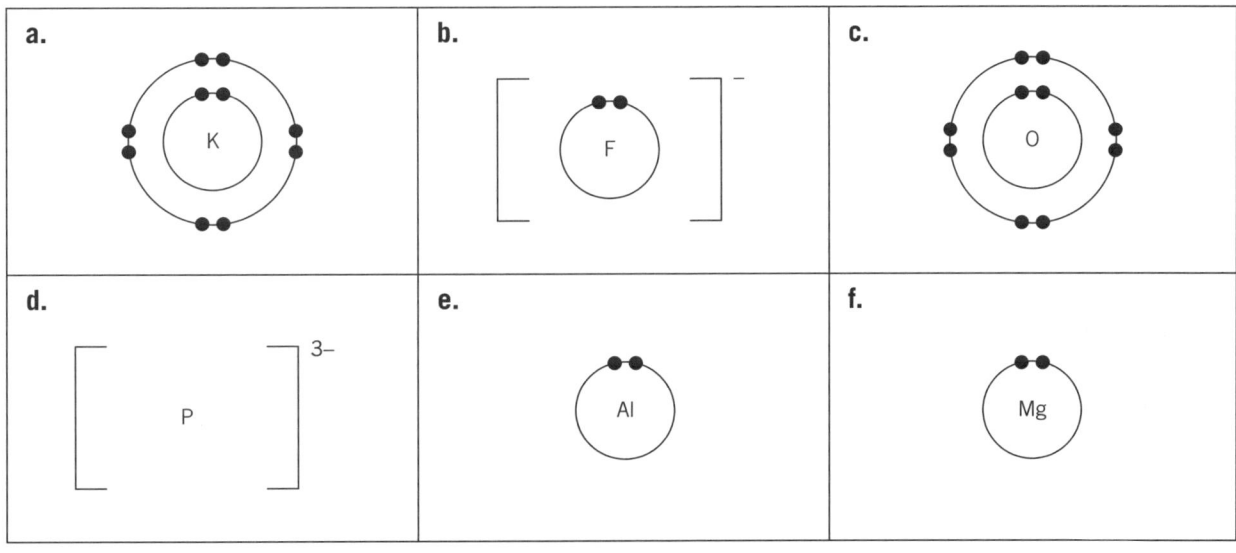

[6]

3. Complete the ionic structure of magnesium sulfide. Show all the electrons as dots.

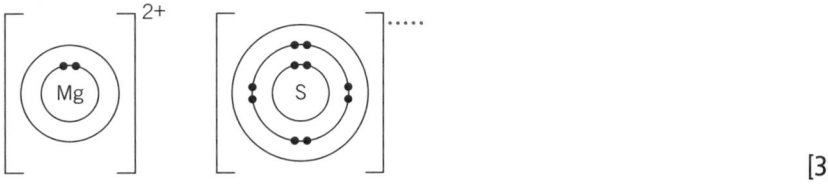

[3]

Link the words **A** to **D** on the left with the correct descriptions **1** to **4** on the right.

A Molecule	**1** A bond formed by sharing a pair of electrons
B Covalent bond	**2** A pair of electrons which is not involved in bonding
C Ionic bond	**3** A particle made of two or more atoms held together by covalent bonds
D Lone pair	**4** The electrostatic attraction between positive and negative ions

[2]

1. Put a ring around the molecules which are diatomic.

CO Cl_2 N_2 N_2O_4 O_2 O_3 P_4 S_8 [2]

2. a. Draw diagrams to show the electron configuration of each of these molecules. Show only the outer shell electrons.

i. hydrogen	**ii.** bromine
iii. hydrogen bromide	**iv.** water

[4]

b. i. How many electrons are there in the outer shell of each atom?

.... [1]

ii. What is the significance of this number of electrons?

....

.... [2]

Complete the passage about covalent molecules using words from the list.

attraction electrons nucleus strong two

A covalent bond is formed when .... atoms combine. It forms because of

the .... force of .... between the .... of one atom

and the outer .... of the atom next to it. [5]

1. Draw dot-and-cross diagram to show the electron configuration of each of these covalent molecules. Show only the outer shell electrons.

oxygen, O_2	nitrogen, N_2
ammonia, NH_3	hydrogen sulfide, H_2S
carbon dioxide, CO_2	ethene, C_2H_4

[6]

Language lab

Complete the passage about ionic structures and covalent structures using words from the list.

attraction   high   intermolecular   ions   low   strong   weak

Simple covalent compounds have .... melting points because the ....

attractive forces are .... Ionic compounds have .... melting points because of

the .... forces of .... between the positive and negative ....

[7]

1. The table gives some properties of some simple molecular covalent compounds and ionic compounds. Complete the table by writing either 'covalent' or 'ionic' in the last column.

compound	melting point / °C	solubility in water	electrical conductivity when molten	simple covalent or ionic?
magnesium oxide	2852	soluble	conducts	
carbon tetrachloride	−23	insoluble	does not conduct	
potassium bromide	734	soluble	conducts	
carbon disulfide	−111	insoluble	does not conduct	
octane	−57	insoluble	does not conduct	

[2]

2. Link the properties **A** to **G** on the left with the correct reasons **1** to **7** on the right.

A Simple molecular compounds have low melting points	**1** ..... because there are no ions or electrons present to conduct.
B Ionic compounds have high melting points	**2** ..... because the molecules cannot form strong enough intermolecular forces with water molecules.
C Simple molecular compounds do not conduct electricity	**3** ..... because they can form relatively strong bonds with the water molecules.
D Some simple molecular compounds do not dissolve in water	**4** ..... because the forces of attraction between the molecules are low.
E Ionic compounds conduct electricity when molten	**5** ..... because they can form relatively strong intermolecular forces with solvent molecules.
F Many ionic compounds dissolve in water	**6** ..... because the ions are free to move.
G Some molecular compounds dissolve in organic solvents	**7** ..... because there are strong forces of attraction between all the ions.

[3]

20

Search for five words which refer to diamond or graphite or both. Words may go downwards or forwards but not backwards, upwards or diagonally. Put your answers in the space below.

C	O	V	A	L	E	N	T	X	U	D
T	E	T	R	A	H	E	D	R	A	L
P	L	U	T	Y	O	C	R	A	C	L
O	I	D	K	E	Z	P	E	G	A	S
G	A	N	O	R	E	G	I	A	N	T
G	A	D	F	S	Z	S	A	N	E	R
I	P	O	S	S	X	U	M	E	X	O
A	R	R	I	S	B	I	G	I	A	P
D	E	L	O	C	A	L	I	S	E	D

[5]

1. a. Some physical properties, **A** to **F** are shown below.

   A conducts electricity   **B** poor electrical conductor   **C** hard

   D high melting point   **E** low melting point   **F** soft

   Write the letters of the properties which refer to:

   Diamond .... [1]

   Graphite .... [1]

   Silicon dioxide .... [1]

   b. Explain by referring to its physical properties and structure why graphite is used as a lubricant.

   .... [3]

2. Link the observations **A** to **E** on the left with the explanations **1** to **5** on the right.

A Giant covalent structures have a high melting point	**1** ..... because the delocalised electrons are free to move along the layers.
B Graphite conducts electricity	**2** ..... because the weak forces between the layers can easily be overcome.
C Diamond is a poor conductor conduct of electricity	**3** ..... because the carbon atoms are packed closer to each other on average.
D Graphite is soft	**4** ..... because it takes a lot of energy to break the large number of strong bonds.
E Diamond is denser than graphite	**5** ..... because all its electrons are involved in covalent bonding.

[2]

3. Describe the arrangement of the atoms in

   Diamond ....   Graphite ....   [2]

Complete the passage about metallic structure and bonding using words from the list.

delocalised   electrons   ionise   layers   positive   voltage

Atoms of metallic elements are generally arranged in closely-packed .... The

metal atoms tend to .... and a mobile 'sea' of .... forms between the

.... metal ions. When a .... is applied to the metal these ....

electrons are able to move.   [6]

1. a. Complete the diagram below to show the structure of a metal. Label your diagram.

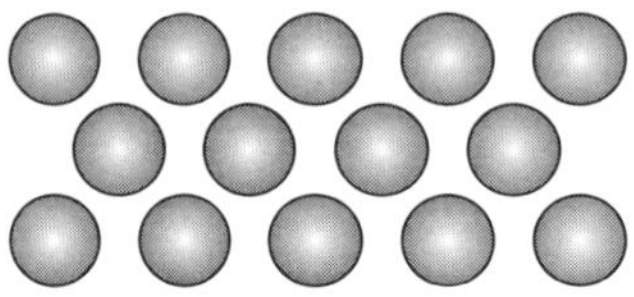

[4]

   b. Use the information in your diagram to explain why metals are ductile.

   ....

   ....

   .... [3]

2. The bar chart shows the melting point of six successive elements **A** to **F** in the Periodic Table.

   a. One of these elements is a giant covalent structure. Which

      one? .... [1]

   b. Which elements are metals? Give a reason for your
      answer.

      ....

      ....

      .... [2]

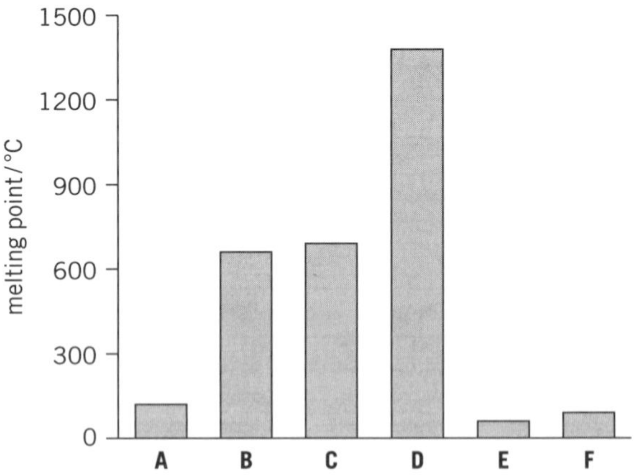

   c. Which elements are non-metals? Give a reason for your answer.

   ....

   .... [2]

Language lab

The element silicon has the symbol Si. But there are other elements hidden there as well if we ignore the lower case (small) letters! For example Li is lithium, C is carbon and O is oxygen.

How many hidden elements can you find in the word arsenic (apart from As)? Find six of these and write their symbols and names. For elements with more than one letter, the letters must be in the same order as the letters in arsenic.

.... [6]

1. **a.** Complete the table to show the combining powers of the atoms shown. Three (H, Li and O) have been done for you.

			H 1								
Li 1						B	C	N	O 2	F	Ne
Na	Mg					Al			S	Cl	
K	Ca		transition elements			Zn					Br

[11]

b. Which atoms in the table lose electrons when they form compounds?

.... [1]

c. Which atoms in the table gain electrons when they form ions?

.... [1]

d. Write the formulae of the following compounds by balancing the combining powers.

i. A compound of H and S ....

ii. A compound of B and O ....

iii. A compound of C and S ....

iv. A compound of C and Br ....

v. A compound of Ca and N ....

vi. A compound of Al and O .... [6]

2. Name these compounds.

A $MgCl_2$ .... [1]

B $Sr(OH)_2$ .... [1]

C $FeSO_4$ .... [1]

D $Zn(NO_3)_2$ .... [1]

E $(NH_4)_2CO_3$ .... [1]

F $Ca(HCO_3)_2$ .... [1]

Link the words **A** to **C** on the left with the correct descriptions **1** to **3** on the right.

A	Displayed formula

1	Shows all of the atoms and all of the bonds

B	Empirical formula

2	Shows the number and type of atoms in a compound

C	Molecular formula

3	Shows the simplest whole number ratio of each type of atom in a compound

[2]

1. Complete these sentences:

    a. Non-metal atoms form non-metal ions by the .... of electrons.   [1]

    b. Metal atoms form metal ions by ....   [1]

2. a. Work out the formulae of compounds **A** to **H** using the list of ions below.

    Al^{3+} Br^- Ca^{2+} Cl^- Fe^{3+} H^+ K^+ Mg^{2+} N^{3-} Na^+ O^{2-} S^{2-}

    A   magnesium bromide ....   **B**   sodium oxide ....

    C   hydrogen chloride ....   **D**   aluminium chloride ....

    E   potassium nitride ....   **F**   calcium sulfide ....

    G   aluminium sulfide ....   **H**   iron(III) oxide ....   [8]

    b. Work out the formulae of compounds **J** to **Q**. Use the list of compound ions below to help you.

    CO_3^{2-} HCO_3^- NH_4^+ NO_3^- OH^- SO_4^{2-}

    J   magnesium nitrate ....   [1]

    K   potassium hydroxide ....   [1]

    L   calcium hydrogencarbonate ....   [1]

    M   ammonium sulfate ....   [1]

    N   calcium hydroxide ....   [1]

    O   aluminium nitrate ....   [1]

    P   lithium carbonate ....   [1]

Language lab

Link the words **A** to **D** on the left with the correct descriptions **1** to **4** on the right.

A Diatomic	**1** Containing one atom
B Monatomic	**2** Substances that are formed in a reaction
C Products	**3** Containing two atoms
D Reactants	**4** Substances that are used up in a reaction [2]

1. Write chemical equations for these 'model' reactions.

   a. $O=O$ and
   $$H-H \qquad H-O-H$$
   gives
   $$H-H \qquad H-O-H$$

   .... [2]

   b. C   and   C   and $O=O$ gives $C\equiv O$ and $C\equiv O$

   .... [2]

2. Balance these equations

   a. $K + Br_2 \rightarrow KBr$

   .... [1]

   b. $Al + O_2 \rightarrow Al_2O_3$

   .... [1]

   c. $Na + O_2 \rightarrow Na_2O$

   .... [1]

   d. $N_2 + H_2 \rightarrow NH_3$

   .... [1]

   e. $Rb + H_2O \rightarrow RbOH + H_2$

   .... [1]

Link the words **A** to **D** on the left with the correct descriptions **1** to **4** on the right.

A Aqueous ion	**1** A letter used to tell you if a substance is solid, liquid, gas, or aqueous
B Precipitate	**2** An ion which does not take part in a reaction
C Spectator ion	**3** A solid formed when two solutions react
D State symbol	**4** An ion dissolved in water

[2]

1. Write down the formulae of the ions present in each of these compounds.

   a. NaOH .... and .... [1]   **b.** $MgCl_2$ .... and .... [1]

   c. $Ba(NO_3)_2$ .... and .... [1]   **d.** $CuSO_4$ .... and .... [1]

   e. Al_2O_3 .... and .... [1]   **f.** $Fe(OH)_2$ .... and .... [1]

2. Write ionic equations for these reactions. In each case, cancel the spectator ions.

   The first one has been partly done for you. Where a solid or molecule is formed, do not separate into ions.
   (Note: solids/molecules have not been shown in the ions/cancel steps.)

   a. $CuCl_2(aq)$ + $2NaOH(aq)$ → $Cu(OH)_2(s)$ + $2NaCl(aq)$

   ions .... + ... $2Na^+ + 2OH^-$ → $Cu(OH)_2(s)$ $2Na^+ + 2$ .... [2]

   cancel .... + ⋯⋯⋯ 2̶N̶a̶⁺̶ ̶+̶ ̶2̶O̶H̶⁻̶   → $Cu(OH)_2(s)$ 2̶N̶a̶⁺̶ ̶+̶ ̶2̶⋯⋯⋯⋯ [1]

   ionic equation $Cu^{2+}(aq)$ + ... (aq) → .... (s) [1]

   b. $BaCl_2(aq)$ + $MgSO_4(aq)$ → $BaSO_4(s)$ + $MgSO_4(aq)$

   ions   .... + ... + ... + ... [2]

   cancel   .... + ... + ... + ... [2]

   ionic equation .... [1]

   c. $Cl_2(aq)$ + $2KI(aq)$ → $I_2(aq)$ + $2KCl(aq)$

   ions   .... + ... + ... [2]

   ionic equation .... [1]

Complete this sentence about relative atomic mass using words from the list.

A_r average   carbon-12   isotopes   twelfth

Relative atomic mass (symbol ....) is the .... mass of the .... of an

element compared to one-................... of the mass of an atom of ....   [5]

1. **a.** Complete the table to calculate the relative molecular mass or relative formula mass of the compounds shown.

compound	number of each atom	A_r of atom	M_r calculation
phosphorus trichloride PCl_3	P = Cl =	P = 31 Cl = 35.5	1×31 $3 \times$ ...... _____ $M_r =$
magnesium hydroxide $Mg(OH)_2$		Mg = 24 O = 16 H = 1	_____ $M_r =$
ethanol C_2H_5OH		C = 12 H = 1 O = 16	_____ $M_r =$
ammonium sulfate $(NH_4)_2SO_4$		N = 14 H = 1 S = 32 O = 16	_____ $M_r =$
glucose $C_6H_{12}O_6$		C = 12 H = 1 O = 16	_____ $M_r =$

[10]

b. Calculate the relative formula mass of these compounds.

i. $Al_2(SO_4)_3$ .... [1]

ii. $Co(NO_3)_2$ .... [1]

iii. $Cr(CO)_6$ .... [1]

iv. $CoCl_2 \cdot 6H_2O$ .... [1]

Language lab

Link the words **A** to **D** on the left with the correct descriptions **1** to **4** on the right.

A Avogadro constant		**1** The average mass of the isotopes of an element compared with 1/12th of the mass of an atom of carbon-12	
B Molar mass		**2** The number of atoms, ions or molecules in a mole of atoms, ions or molecules	
C Mole		**3** The mass of a mole of substance in grams	
D Relative atomic mass		**4** The amount of substance that that has the Avogadro number of particles	

[2]

1. Complete the table using the A_r values below.

   C = 12, Ca = 40, H = 1, O = 16, P = 31, S = 32

element or compound	formula mass, M_r	mass taken / g	number of moles
O_2	32	4	
NaCl		11.7	
$CaSO_4$		27.2	
P_2O_5			0.4
CO_2			0.1
P_4		86.8	
CH_4			24.0

[13]

2. Calculate the number of molecules of ethene, C_2H_4, in 560 g of ethene by following the steps below.

   a. Relative molecular mass of ethene = ... [1]

   b. Moles of ethene = ... [1]

   c. Number of molecules of ethene (Avogadro constant = 6.02×10^{23}) =

   .... [1]

3. 0.2 moles of aluminium has a mass of 5.4 g.

   Calculate the molar mass of aluminium.   [3]

Complete these sentences about mole calculations using words from the list.

adding   atomic   atoms   dividing   moles   number   relative

To find the .... of moles of a compound, we need to know the mass of compound taken

and the .... molecular mass of the compound. The relative molecular mass is

found by .... together the relative .... masses of all the .... (or ions)

in the compound. The number of .... is found by .... the mass of compound

taken by the relative molecular mass. [7]

1. Use simple proportion to do these calculations about the reacting masses.

   When 48 g of magnesium are burnt completely in oxygen, 80 g of magnesium oxide is formed.

   $$2Mg \;+\; O_2 \;\rightarrow\; 2MgO$$

   a. Complete the calculation to show the mass of magnesium oxide formed when 12 g of Mg are burnt?

   $$\frac{..............}{..............} \times 80 \;=\; .... \, g$$ [2]

   b. What mass of magnesium is needed to form 8 g of magnesium oxide?

   .... [1]

   c. What mass of magnesium oxide is formed when 168 g of magnesium are burnt?

   .... [1]

2. a. Complete the calculation to show the percentage by mass of carbon in ethane, C_2H_6.

   $$\frac{2 \times ....}{(...... \times 12) + (...... \times 1)} \times 100 = ... \, \%$$ [2]

   b. Calculate the percentage by mass of nitrogen in ammonia, NH_3. Show your working.

   .... % [2]

   c. Calculate the percentage by mass of calcium in calcium carbonate, $CaCO_3$. Show your working.

   .... % [2]

29

Language lab

Link the words **A** to **D** on the left with the correct descriptions **1** to **4** on the right.

| A Excess reactant | 1 The reactant that is not in excess so is completely used up in the reaction |

| B Products | 2 The ratio of the atoms of each reactant and product in the chemical equation |

| C Limiting reactant | 3 The substances formed as result of a chemical reaction |

| D Stoichiometry | 4 The reactant that is left over after the reaction is complete |

[2]

1. Complete this equation to show the reacting masses.

$$4PH_3 \quad \rightarrow \quad P_4 \quad + \quad 6H_2$$

$4 \times$ .... $\rightarrow$ $4 \times$ .... $+$ $6 \times$ ....   [1]

.... g $\rightarrow$ .... g $+$ .... g   [1]

2. Use the number of moles shown in the equation below to answer the questions which follow.

$$I_2O_5 \; + \; 5CO \; \rightarrow \; I_2 \; + \; 5CO_2$$

M_r *values*   414   140   254   220

   a. How many moles of CO react with 1 mole of I_2O_5? ....   [1]

   b. How many moles of CO_2 are formed from 1 mole of CO? ....   [1]

   c. How many moles of I_2 are formed using 10 moles of CO? ....   [1]

   d. What mass of CO_2 is formed using 24.84 g of I_2O_5 and excess CO?

     .... [2]

   e. What mass of I_2 is formed using 21 g of CO and excess I_2O_5?

     .... [2]

3. 168 g of iron reacts with excess oxygen to form 232 g of an oxide of iron.

   A_r values: Fe = 56; O = 16

   a. Calculate the moles of iron in the iron oxide .... [1]

   b. Calculate the moles of oxygen in the iron oxide .... [2]

   c. Deduce the formula of this oxide of iron .... [2]

Complete these sentences about gas volumes using words from the list.

conditions   dm³   mole   molecules   number   oxygen   temperature

At room .... and pressure, one .... of any gas occupies 24 ....

(24000 cm³). Because there is the same .... of moles, there is also the same number of

gas .... in a given volume. So, 50 cm³ of chlorine and 50 cm³ of .... under

the same .... contain the same number of molecules.   [7]

1. Complete the table to show the mass, moles or volume of different gases.

gas	M_r of gas	mass of gas / g	moles of gas / mol	volume of gas / dm³
ammonia	17	8.5		
oxygen	32			48
carbon dioxide	44	3.08		
hydrogen chloride		292	8	
ethane	30			3

[10]

2. 48 cm³ of hydrogen is mixed with 25 cm³ of oxygen at r.t.p. The mixture reacts when a spark is applied and water is formed.

$$2H_2(g) + O_2(g) \rightarrow 2H_2O(l)$$

a. Calculate the number moles of hydrogen and oxygen at the start of the reaction.

[2]

b. Use the stoichiometry of the reaction to deduce which reactant is limiting.

[2]

c. Calculate the mass of water formed in the reaction. Show all your working.

[2]

31

Complete this relationship using words from the list.

impure   mass   percentage   pure

.... purity = ... of .... product × 100   [4]

mass of .... product

1. 18 g of limestone was reacted with excess dilute hydrochloric acid. 3840 cm³ of carbon dioxide were formed at r.t.p. Work through the calculation to find the percentage purity of an impure sample of limestone, calcium carbonate, $CaCO_3$.

   Give your answer to 2 significant figures. A_r values: C = 12, Ca = 40, O = 16

$$CaCO_3(s) + 2HCl(aq) \rightarrow CaCl_2(aq) + CO_2(g) + H_2O(l)$$

   a. Molar mass of calcium carbonate = ... g / mol   [1]

   b. Volume of CO_2 in dm³ = ... dm³   [1]

   c. Moles of CO_2 = ... = ... mol   [1]

   ....

   d. Moles of $CaCO_3$ in impure limestone = ... mol   [1]

   e. Mass of $CaCO_3$ in impure limestone = ... g   [1]

   f. Percentage purity = ...%   [1]

2. Methyl benzoate can be prepared by reacting methanol with benzoic acid.

$$CH_3OH + C_6H_5CO_2H \rightarrow C_6H_5CO_2CH_3 + H_2O$$

   methanol   benzoic acid   methyl benzoate

   When 24.4 g of benzoic acid is reacted with excess methanol, 25.84 g of methyl benzoate is produced. Calculate the percentage yield of methyl benzoate. A_r values: C = 12, H = 1, O = 16

   a. Molar mass of benzoic acid = ... g / mol   [1]

   b. Moles of benzoic acid = ... mol   [1]

   = ... = ...

   ....

   c. Moles of methyl benzoate expected (if 100 % yield) = ... mol   [1]

   d. Molar mass of methyl benzoate = ... g / mol   [1]

   e. Mass of methyl benzoate expected (if 100 % yield) = ... g   [1]

   f. Percentage yield = ...%   [1]

Language lab

Complete these sentences about empirical and molecular formulae using words from the list.

atoms　　combine　　compound　　each　　molecule　　simplest

The molecular formula of a .... shows the number of .... type of atom in one

.... The empirical formula shows the .... ratio of .... which

....

[6]

1. Complete the following calculations to find the empirical formulae.

   A compound of lead and chlorine contains 20.7 g of lead and 14.2 g of chlorine.

   A_r values: Pb = 207, Cl = 35.5

   moles of Pb = ... mol　　　moles of Cl = ... mol　　　[1]

   Divide by　　　Pb　　......................... 　　　　Cl　　.........................
   lowest number
   of moles　　　　　　......................... 　　　　　　　　......................... 　　　[1]

   Result of division = 　......................... 　　　　　 = ...

   Simplest ratio .... So empirical formula is .... 　　　[2]

2. Write the empirical formula of the following compounds whose molecular formula has been given.

   a. Hydrogen peroxide, H_2O_2.　　Empirical formula .... [1]

   b. Antimony(III) oxide, Sb_4O_6.　Empirical formula .... [1]

   c. Butane, C_4H_{10}.　　　　　　Empirical formula .... [1]

3. Complete the table to deduce the empirical formula mass and molecular formulae of compounds **A** to **C**.
   A_r values: C = 12, Cl = 35.5, H = 1, O = 16, P = 31, S = 32

empirical formula	empirical formula mass / g	relative molecular mass, M_r / g	molecular formula
A P_2O_3		220	
B SCl		135	
C CH_2O		60	

[6]

Language lab

Complete this relationship using words from the list.

amount   dm³   moles   solute   volume

concentration in .... per dm³ = ... of .... in moles   [5]

.... in ....

1. Complete the table to show the moles and mass of solute and the concentration.

solute	M_r of solute	mass of solute / g	volume of solution cm³ or dm³	concentration of solution mol / dm³
sodium hydroxide	40	8	250 cm³	
silver nitrate	170		200 cm³	0.5
copper(II)sulfate	160	40		0.125

[3]

2. Work through this calculation to find the concentration of a solution of sodium hydroxide when 25.0 cm³ of a solution of sodium hydroxide is exactly neutralised by 12.2 cm³ of sulfuric acid of concentration 0.100 mol / dm³.

a. moles of acid = ... × $\dfrac{........................}{1000}$ = ... mol H_2SO_4 [1]

b. The equation for the reaction is: $2NaOH + H_2SO_4 \rightarrow Na_2SO_4 + 2H_2O$

i. How many moles of NaOH react with 1 mole of H_2SO_4? ....   [1]

ii. How many moles of NaOH are needed to react with the amount (in mol) of H_2SO_4 you calculated in part a?

.... [1]

c. Calculate the concentration of NaOH in 25 cm³ of the sodium hydroxide solution.

[2]

Use the clues below to do this crossword

Across: 1. Break down

           5. Conducts current in or out of the electrolyte

           6. A charged atom or group of atoms

           7. A power source

Down: 2. Negative electrode

           3. Positive electrode in reverse

           4. Electrodes + electrolyte make an electrochemical ....

[7]

1. Explain why aluminium cables with a steel core are used for overhead power cables. In your answer refer to the physical properties of both metals.

....

.... [3]

2. Complete the diagram of an electrolysis cell by labelling the anode, the cathode, the electrolyte and the direction of current flow in the external circuit.

[4]

3. What is meant by these terms?

   a. Electrolysis ....

   .... [2]

   b. Electrolyte .... [1]

   c. Anode .... [1]

4. a. Give two properties that an electrode should have.

   1 .... [1]

   2 .... [1]

   b. Name two elements that are commonly used as electrodes.

   .... and .... [2]

Complete these sentences about the electrolysis of dilute sulfuric acid using words from the list.

anode   bubbles   cathode   hydroxide   ions   platinum   water

When dilute sulfuric is electrolysed using .... electrodes we observe .... at each

electrode. Hydrogen is formed at the .... and oxygen is produced at the .... We

can think of this electrolysis being the electrolysis of .... The hydrogen is formed from

hydrogen .... and the oxygen from .... ions.   [7]

1. Complete the table to show the electrode products and observations at the anode when various substances are electrolysed using graphite electrodes.

electrolyte	cathode (–) product	anode (+) product	observations at the anode
Molten zinc bromide			
Molten magnesium chloride			
Molten calcium oxide			
Molten lead(II) iodide			

[12]

2. Complete these equations for the electrolysis of these molten electrolytes.

   a. ....NaCl(l) $\rightarrow$ ....Na(l) + Cl_2(.....)   [3]

   b. 2MgO(l) $\rightarrow$ 2 ....(l) + ... (g)   [2]

   c. .... (.....) $\rightarrow$ Pb(l) + I_2(g)   [2]

3. When hydrochloric acid is electrolysed, oxygen is formed at the anode and hydrogen is formed at the cathode. Complete the diagram in the space on the right to show:

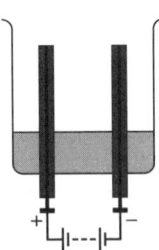

   a. how the gases are collected.   [2]

   b. labels for the anode and cathode.   [1]

   c. a label to the electrolyte.   [2]

Join up these fragments to form a sentence describing why hydrogen is sometimes formed when aqueous solutions are electrolysed.

.........hydrogen ions in water, ...is more reactive than hydrogen, ...   If a metal...................

.........hydrogen arising from.........   .... bubbles off.   ....the metal ions stay in solution and ....

....

.... [2]

1. Complete the table to show the electrode products and observations at the anode when various substances are electrolysed using graphite electrodes.

electrolyte	cathode (–) product	anode (+) product	observations at the anode
concentrated KCl(aq)			
dilute H_2SO_4(aq)			
dilute NaCl(aq)			
concentrated HCl(aq)			
dilute KI(aq)			

[15]

2. The diagram shows a cell used for electrolysing brine (concentrated sodium chloride).

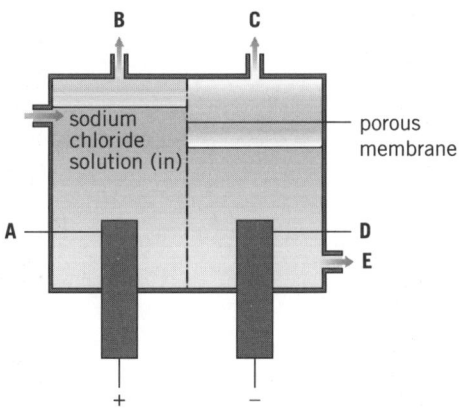

a. Give the names of the gases collected at

   B .... C..................................... [2]

b. Which letter in the diagram represents the anode? .... [1]

c. Explain why oxygen is not given off at the anode.

   ....

   .... [2]

Language lab

Complete these sentences about electrolysis using words from the list.

anode   cathode   electrons   lose   move   negative   oxidation   reduction

During electrolysis, positive ions move towards the .... where they gain .... This is a

.... reaction. Negative ions .... to the .... where they .... electrons.

This is an .... reaction. In the external circuit the electrons travel in the wires from the ....

pole of the battery to the cathode.   [8]

1. Complete these diagrams to show

   • The movement of ions during electrolysis by drawing arrows

   • What happens to the ions in terms of electron loss or gain to or from the electrodes

     (Show this by curly arrows)

a.   molten zinc bromide

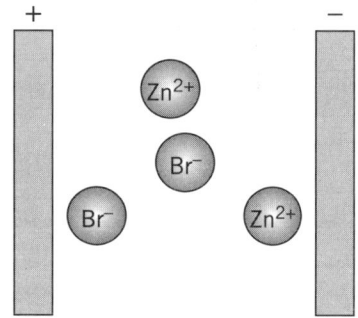

b.   dilute sulfuric acid

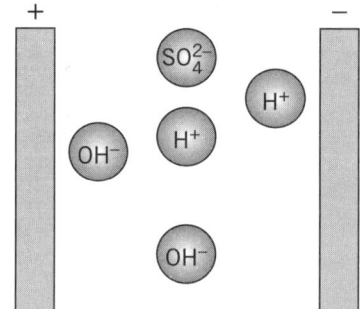

c.   concentrated hydrochloric acid

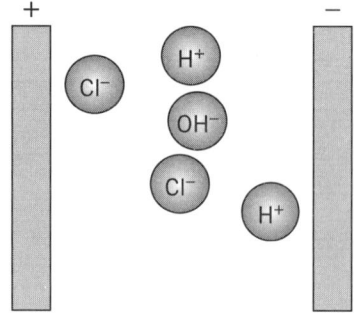

d.   aqueous copper(II) sulfate

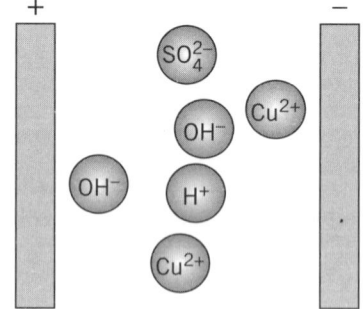

[8]

2. Complete these half equations for the reaction at the electrodes

   a. $Zn^{2+} + \ldots\ldots \rightarrow Zn$ [1]

   b. $\ldots\ldots Cl^- \rightarrow \ldots\ldots + \ldots\ldots$ [2]

   c. $\ldots\ldots H^+ + \ldots\ldots \rightarrow \ldots\ldots$ [2]

   d. $Al^{3+} + \ldots\ldots \rightarrow \ldots\ldots$ [1]

   e. $\ldots\ldots OH^- \rightarrow \ldots\ldots + \ldots\ldots H_2O + \ldots\ldots$ [2]

Language lab

Complete these sentences about the purification of copper using words from the list.

anode   atoms   cathode   copper   electrolyte   gain   impure   ions   solution

The electrolysis cell has an .... strip of copper as the .... and a pure strip of

.... as the cathode. The .... is a solution of copper(II) sulfate. At the anode,

copper .... lose electrons and go into .... as copper(II) .... At the

cathode copper(II) ions .... electrons and are deposited on the .... as

copper atoms.   [9]

1. **a.** Complete these equations for the reactions at the electrodes when copper is refined using copper electrodes dipping into aqueous copper(II) sulfate.

   Anode: $Cu \rightarrow$ .... + ... e^- [2]

   Cathode: .... + ... $\rightarrow Cu$ [2]

   b. Explain why the aqueous copper(II) sulfate does not get lighter in colour during this electrolysis.

   ....

   .... [2]

2. Copper(II) sulfate can be electrolysed using graphite electrodes or copper electrodes. Complete the table to show what happens at each electrode and to the electrolyte.

What happens....	using graphite electrodes	using copper electrodes
to the mass of the electrodes	anode: cathode:	anode: cathode:
to the appearance of the electrodes during electrolysis	anode: cathode:	anode: cathode:
to the electrolyte (Give any observations)		

[10]

Complete these sentences about electroplating using words from the list.

cathode   electrolyte   electroplated   ions   metal   negative   plating   pole   power

The object to be .... is connected to the .... pole of the ....

supply. The object becomes the .... A strip of the plating .... is connected

to the positive .... of the power supply. The .... metal is the anode. The

.... is a solution containing .... of the plating metal.   [9]

1. A nickel jug can be electroplated with silver. Look at the diagram on the right and then answer the questions.

    a. On the diagram label the anode, **A**, the cathode, **C**, and the electrolyte, **E**.

    [2]

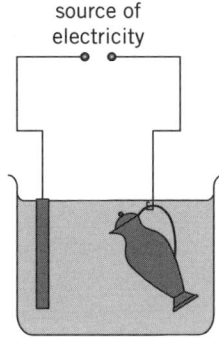

source of electricity

    b. What happens to the cathode as the electroplating proceeds.

    ....

    ....

    ....

    .... [2]

    c. Complete these equations for the reactions at the anode and cathode.

    Anode: $Ag \rightarrow$ .... + ...   [2]

    Cathode: .... + ... $\rightarrow$ ....   [2]

2. a. Explain why plating an iron object with tin prevents the iron from rusting.

    ....

    .... [2]

    b. Give one other reason why objects are electroplated.

    .... [1]

Language lab

Search for the names of five substances related to the manufacture of aluminium by electrolysis of aluminium oxide. Words may go downwards or forwards but not backwards, upwards or diagonally. Write the names in the space below.

O	C	R	Y	O	L	I	T	E
X	A	L	U	M	I	N	A	R
Y	Z	B	X	X	F	D	A	P
G	R	A	P	H	X	F	E	I
E	P	U	N	O	M	B	R	N
N	A	X	O	C	A	A	O	G
N	F	I	C	A	R	B	O	N
E	R	T	P	L	O	V	D	I
D	Y	E	A	R	A	N	C	I

[5]

1. The diagram shows an electrolysis cell used to extract aluminium from aluminium oxide.

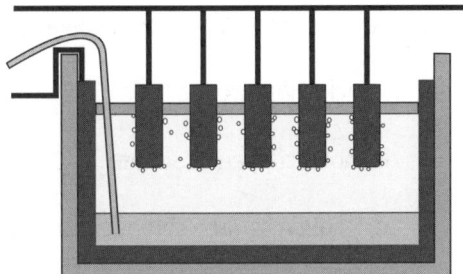

On the diagram above, label the electrolyte as **E**, the cathode as **C**, the anode as **A** and the molten aluminium as **M**.   [4]

2. Explain why the electrolysis mixture is mainly cryolite and calcium fluoride with about 5% aluminium oxide.

....

....

.... [3]

3. a. Complete these equations for the reactions at the anode and cathode.

   i. Al^{3+} + ... $\rightarrow$ ....   [2]

   ii. ....O^{2-} $\rightarrow$ .... + ...   [2]

   b. Construct the overall equation for this electrolysis.

   .... [2]

41

Join up these fragments to form a sentence about ceramics.

.........because they resist the flow ....   Ceramics are useful.........   .... they have very high.........

.... insulators, not only .... melting points.   .... of electricity, but also because.........

....

.... [2]

1. **a. i.** Label the diagram to show the apparatus used to show whether or not a solid conducts electricity. [1]

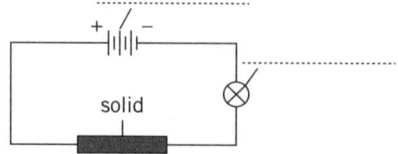

solid

ii. On the diagram above put an arrow to show the direction of flow of the electrons. [1]

2. Which of these substances are good conductors and which are poor conductors?

Write 'good' or 'poor' in the spaces next to each substance.

a. graphite ....   **b.** solid copper(II) sulfate ....

c. plastic ....   **d.** molten sodium chloride ....

e. nickel ....   **f.** ceramics .... [6]

3. Link the phrases **A** to **D** on the left with the phrases **1** to **4** on the right.

A Molten sodium chloride conducts electricity....	**1** ....because delocalised electrons move throughout the structure when a voltage is applied.
B Metals conduct electricity....	**2** ....because the ions are not free to move.
C Sulfur does not conduct electricity......	**3** .... because the ions are free to move.
D Solid sodium chloride does not conduct electricity....	**4** ...... because it does not contain mobile ions or delocalised electrons.

[2]

Link the words **A** to **D** on the left with the best descriptions **1** to **4** on the right.

A	Chemical change

1	Thermal energy is given out

B	Endothermic

2	No new substances are formed

C	Exothermic

3	Thermal energy is absorbed

D	Physical change

4	New substances are formed

[2]

1. Underline the physical changes.

   Burning magnesium in air   Separating iron from sulfur using a magnet

   Rusting of iron   Melting zinc

   Distilling plant oils from a mixture of plant oils and water   [3]

2. Describe these changes as either exothermic or endothermic.

   a. The decomposition of copper(II) carbonate by heating .... [1]

   b. Burning paraffin.. [1]

   c. Your tongue gets cold when you put sherbet on it .... [1]

   d. The temperature of the solution rises when concentrated hydrochloric acid is diluted.

      .... [1]

3. Complete these sentences about exothermic and endothermic reactions using words from the list. Not all the words are used.

   decreases   enthalpy   faster   increases   light   slower   solution   surroundings   thermal

   a. Endothermic reactions absorb thermal energy from the ....

   b. In an exothermic reaction, the temperature of the surroundings ....

   c. The surroundings includes the test tube, the air around the test tube and the .... in the test tube.

   d. Another word for thermal energy transfer is .... change.

   e. 'Heat' is a word that is often used for the transfer of .... energy.

   f. At higher temperatures particles move ....

43

Language lab

Link the words **A** to **D** on the left with the descriptions **1** to **4** on the right.

A Activation energy	**1** The thermal energy change in a chemical reaction
B Enthalpy change	**2** The general progress of a reaction from reactants to products
C Reaction pathway	**3** The energy which relates to the amount of vibration and movement of particles
D Thermal energy	**4** The minimum energy needed for colliding particles to react [2]

1. a. Complete these reaction pathway diagrams, **L** and **M**, for an exothermic and an endothermic reaction. Include an arrow in each diagram in the correct direction.

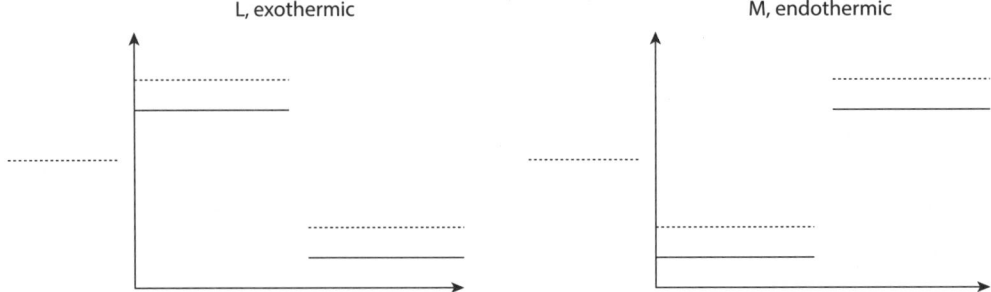

L, exothermic   M, endothermic

[5]

b. Explain why reaction pathway diagram **L** represents an exothermic reaction.

....

.... [2]

c. In the space on the right draw a labelled reaction pathway diagram, including activation energy, for an endothermic reaction. Include the following:

- The axes suitably labelled

- The words 'reactant' and 'products'

- The activation energy as an arrow

- The enthalpy change as an arrow.

[5]

Complete these sentences about bond breaking and bond making during chemical reactions using words from the list.

absorbed　　bonds　　out　　greater　　new　　reactants　　released

In an endothermic reaction, the energy .... in bond breaking is .... than the

energy .... when new .... are formed. In an exothermic reaction, the energy

given .... when .... bonds are formed is greater than the energy taken in

when the bonds in the .... are broken. [7]

1. Complete the table to calculate the energy change when methane reacts with excess oxygen to form carbon dioxide and water.

$$\text{H}-\overset{\displaystyle \text{H}}{\underset{\displaystyle \text{H}}{\text{C}}}-\text{H} \;+\; 2\,\text{O=O} \;\rightarrow\; \text{O=C=O} \;+\; 2\,\text{H}-\text{O}-\text{H}$$

Bond energies in kJ / mol : C–H 413,　O=O 498,　C=O 805, O–H 464

bonds broken (endothermic $^+$) / kJ / mol	bonds formed (exothermic $^-$) / kJ / mol
4 × (C–H) = 4 × 413 = ... × (O=O) = ... = ...	
Total　　　　　　　...............	

Overall energy change = (+ ....) + (– ....) = ... kJ / mol [5]

2. Hydrogen reacts with chlorine to produce hydrogen chloride.

$$\text{H}_2(\text{g}) \;+\; \text{Cl}_2(\text{g}) \;\rightarrow\; 2\text{HCl}(\text{g})$$

Calculate the enthalpy change of this reaction, using the following bond energy values and state whether the reaction is exothermic or endothermic.

Bond energies in kJ / mol: H–H 435.9,　Cl–Cl 243.4,　H–Cl 432.0

..............kJ / mol [4]

Link the energy sources **A** to **C** on the left with the descriptions **1** to **3** on the right.

A Coal	**1** This is largely methane, the simplest hydrocarbon
B Natural gas	**2** A black liquid which is distilled to give a variety of fuels such as gasoline
C Petroleum	**3** A black solid formed from the decay of plants in the absence of oxygen

[2]

1. The table shows the temperature changes produced by burning the fuels **A**, **B**, **C** and **D** underneath a copper can filled with 100 cm³ of water.

fuel	amount burnt / g	initial temperature of water / °C	final temperature of water / °C	density in g / cm³
A	2.0	20	25	0.90
B	1.0	18	21	0.55
C	4.0	21	29	0.87
D	3.5	25	24	1.10

a. Which fuel gave the greatest temperature rise in the experiment? .... [1]

b. Which fuel produced the most energy per gram? .... [1]

c. i. Which fuel would be the cheapest to transport? Give a reason for your answer.

   .... [2]

   ii. Apart from your answer to part **c. i.** give one other factor which would influence how expensive it is to transport a fuel.

   .... [1]

2. Complete these equations for the complete combustion of different compounds.

   a. .... $H_2 + O_2 \rightarrow$ .... [2]

   b. .... $C_2H_6 +$ .... $\rightarrow$ .... + ... [2]

   c. $C_7H_{16} +$ .... $\rightarrow$ .... + ... [2]

   d. .... $H_2S +$ .... $\rightarrow$ .... + ... SO_2 [2]

3. Complete these sentences by using suitable words.

   Methane is burnt in a limited supply of .... The combustion is ....

   Carbon .... is formed as well as small amounts of carbon, carbon dioxide and water. [3]

Complete these sentences about fuel cells using words from the list.

alkali   electrons   external   negative   oxygen   platinum   porous   reactions

A fuel cell consists of two .... electrodes coated with .... The electrolyte is either an

acid or an .... Hydrogen and .... are bubbled through the porous electrodes where

the .... take place. When connected to an .... circuit, ... flow from

the .... electrode to the positive electrode.   [8]

1. The diagram shows a hydrogen–oxygen fuel cell.

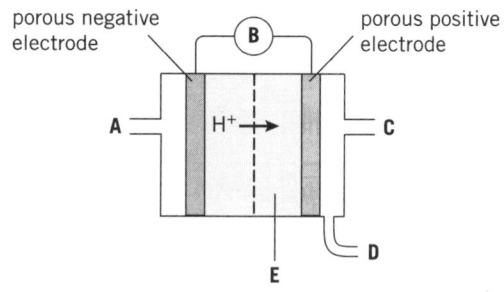

a. Give the name of the instrument labelled **B**. .... [1]

b. The product of the reaction in the fuel cell is water. Which letter **A, C, D** or **E** shows where water is collected?

    .... [1]

c. Which letter **A, C, D** or **E** shows where hydrogen enters the cell?.. [1]

2. Complete these half equations for the reactions taking place in the fuel cell.

a. .... H_2 + ... OH^- → .... H_2O + ...   [2]

b. .... + ... H_2O + ... → .... OH^- [2]

3. Give two advantages and two disadvantages of using fuel cells instead of petrol to power a car.

advantage 1 ....

advantage 2 ....

disadvantage 1 ....

disadvantage 2 .... [4]

Complete these sentences about methods for following the course of a reaction using words from the list.

    decreases   products   quickly   rate   second   time   used   volume

To find the rate of reaction we can either measure how .... the reactants are .... up or

how quickly the .... are formed. To calculate the .... of reaction we need to find out how

some measurement changes with .... For example, the .... of gas given off per

.... , or how the mass of the reaction mixture .... with time.   [8]

1. Put these in order of increasing rate of reaction.

    A Immediate precipitation   B Rusting   C Paint drying   D Magnesium burning in air

    .... [1]

2. The diagrams show the reaction of 50% nitric acid with copper.

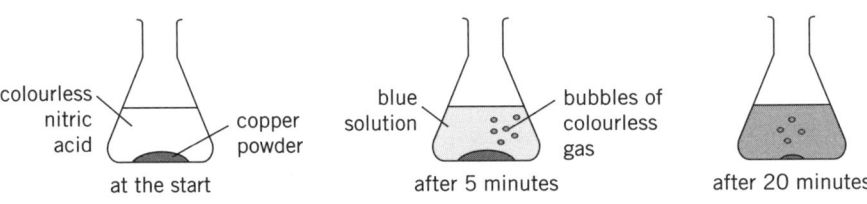

    a. Give three pieces of information from the diagram that show that a chemical reaction is occurring.

        ....
        .... [3]

    b. Suggest three different ways by which you could measure the rate of this reaction.

        1 ....

        2 ....

        3 .... [3]

3. Suggest why the course of this reaction could be monitored by measuring electrical conductivity.

    $$H_2O_2(aq) + 2I^-(aq) + 2H^+(aq) \rightarrow 2H_2O(l) + I_2(aq)$$

    .... [2]

Link the words **A** to **D** on the left with the phrases **1** to **4** on the right.

A	Accuracy

1	This changes in value during the experiment: You measure these values

B	Control variable

2	The variable that you vary the values of in the experiment

C	Dependent variable

3	How close the measurements are to their true value

D	Independent variable

4	This is kept constant throughout the experiment

[2]

1. A student wants to deduce the rate of reaction of hydrochloric acid with magnesium to produce hydrogen.

   The student suggested the apparatus shown. There are several errors in this diagram.

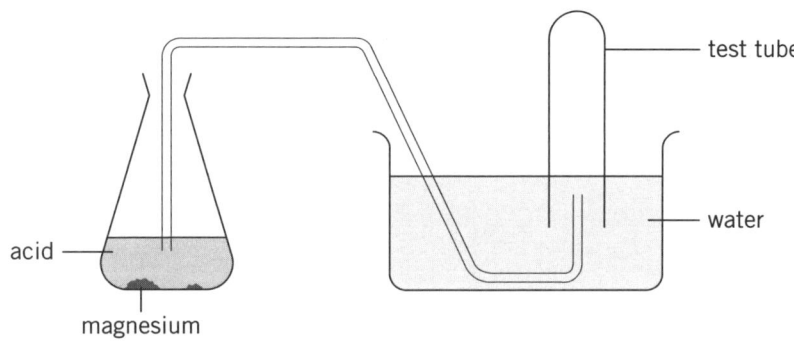

   acid

   magnesium

   test tube

   water

   a. Identify four errors in this diagram that stop the experiment from working correctly. For each error suggest what should be done to improve the experiment.

      1 .... [2]

      2 .... [2]

      3 .... [2]

      4 .... [2]

   b. Name a piece of apparatus, essential to the working of the experiment, that the student has not drawn.

      .... [1]

2. Two ways of determining the progress of this reaction are measuring the increase in gas volume in a graduated syringe or measuring the decrease in mass of the reaction mixture.

   Explain which method is likely to give more accurate results.

   ....

   ....

   .... [2]

Join up these fragments to form two sentences describing how the rate of reaction between zinc and hydrochloric acid changes with time.

...... per second is high .... At the start of the reaction, the....... .... of the graph decreasing.

.... volume of hydrogen given off .... As the reaction proceeds, the volume ....

.... of hydrogen given off per second decreases, ... which is shown by the gradient ....

....

.... [2]

1. The decomposition of hydrogen peroxide is speeded up by catalysts.

$$2H_2O_2(aq) \rightarrow 2H_2O(l) + O_2(g)$$

A student investigated how the rate of reaction changes when two different catalysts are used.

The results using catalyst **A** are shown below.

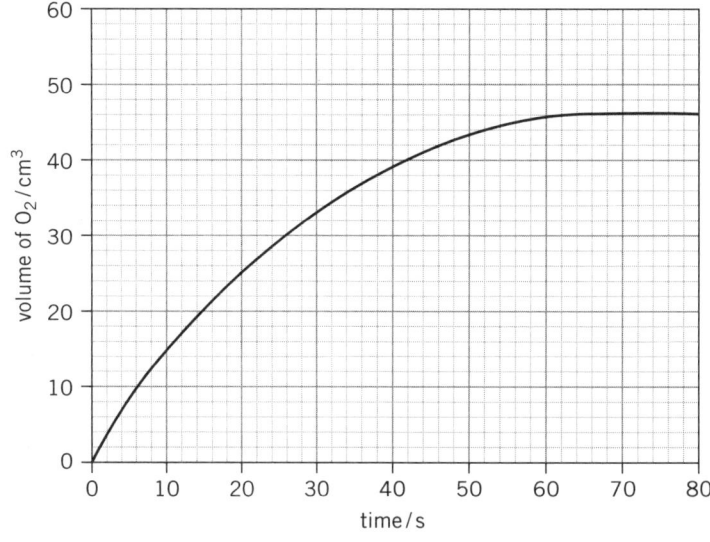

a. At what time is the reaction just complete? .... [1]

b. What volume of gas has been released when the reaction is just complete?

   .... [1]

c. What volume of gas has been produced in the first 30 seconds of the reaction?

   .... [1]

d. The reaction was repeated using catalyst **B**. The results are shown in the table.

time / s	0	4	10	20	30	40	50	60	70	80
volume / cm³	0	20	33	44	50	54	56	59	60	60

Plot a graph of these results on the same grid as for catalyst **A** above. Draw the curve of best fit through the points. [2]

Language lab

Complete these sentences about catalysts using words from the list.

end   increases   lower   rate   reaction   unchanged

A catalyst is a substance which .... the .... of a reaction. The catalyst

remains .... at the .... of the reaction. A catalyst works by providing a route

for the .... which has a .... activation energy.   [6]

1. a. Calculate the surface area of a cube of marble of dimensions 2 cm × 2 cm × 2 cm.

   (If you are not sure how to do this see Unit 23.5.)

   .... [1]

   b. The cube is cut up into 8 cubes of dimensions 1 cm × 1 cm × 1 cm.

   Calculate the total surface area of all these cubes.

   .... [1]

   c. Which set of cubes will reacts faster with hydrochloric acid? Explain your answer.

   .... [2]

2. 4 g of large marble chips reacts with excess dilute hydrochloric acid

$$CaCO_3(s) \quad + \quad 2HCl(aq) \quad \rightarrow \quad CaCl_2(aq) \quad + \quad CO_2(g) \quad + \quad H_2O(l)$$

The experiment was repeated with 4 g of medium-sized marble chips then with 4 g of small marble chips. All other conditions stay the same.

a. On the axes below draw a sketch graph to show how the volume of carbon dioxide released changes with time using large, **L**, medium, **M**, and small, **S**, marble chips. Label your lines **L**, **M** and **S**.

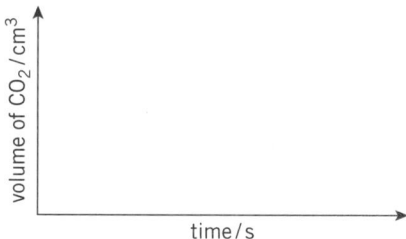

[2]

b. The experiment was repeated but this time measuring the decrease in the mass of the reaction mixture with time. In the space on the right, draw a sketch graph to show how the mass of the reaction mixture changes with time for the small and the large marble chips.

[4]

Language lab

Complete these sentences about the collision theory using words from the list.

bonds   collide   energy   frequency   increases   rate

In order to react, particles must .... with each other. The collisions must have enough

.... to break .... to allow a reaction to happen. Increasing the concentration of

a reactant .... the .... of collisions and so increases the .... of

reaction.   [6]

1. Use the particle diagrams below to answer the following questions about the reaction

$$Mg(s) + 2HCl(aq) \rightarrow MgCl_2(aq) + H_2O(l)$$

   a. Complete the diagram on the right to show the particles of acid and water in a concentrated solution of acid.

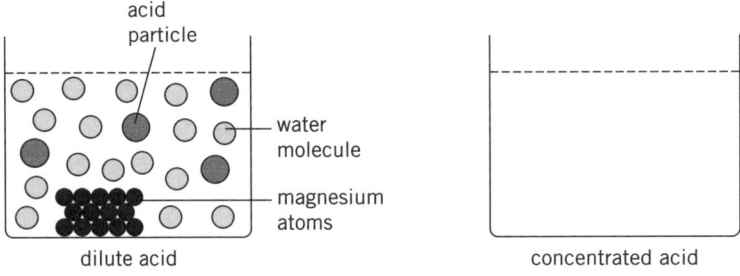

acid particle — water molecule — magnesium atoms

dilute acid   concentrated acid

                                                                                                                            [3]

   b. Complete the diagram on the right to show the relative number of acid, magnesium and water particles when the reaction is nearly complete.

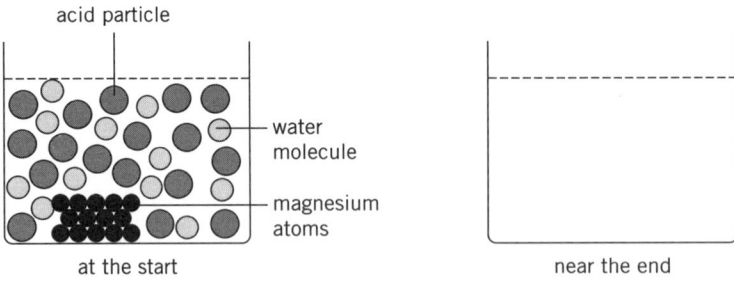

acid particle — water molecule — magnesium atoms

at the start   near the end

                                                                                                                            [3]

2. The reaction shown below takes place in a closed container. Explain using ideas about colliding particles why the rate increases when the pressure increases.

$$2SO_2(g) + O_2(g) \rightarrow 2SO_3(g)$$

....

.... [2]

Language lab

Link the words **A** to **D** on the left with the phrases **1** to **4** on the right.

| A   Activation energy |
| A   Effective collisions |

A Activation energy

B Effective collisions

C Frequency of collisions

D Kinetic energy

1 The number of collisions in a given time

2 The energy associated with moving particles

3 The minimum energy particles must have when they collide in order to react

4 Collisions which result in a reaction taking place

[2]

1. A student investigated how increasing the temperature affects the rate of the reaction of magnesium with 1.0 mol / dm³ hydrochloric acid. The student measured the volume of hydrogen given off in 1 minute at seven different temperatures. The table shows the results.

temperature / °C	20	30	41	45	50	55	60
volume of H$_2$ / cm³	10	20	44	56	80	110	160

    a. Plot a graph of these results on the grid below. Draw the best curve through the points.

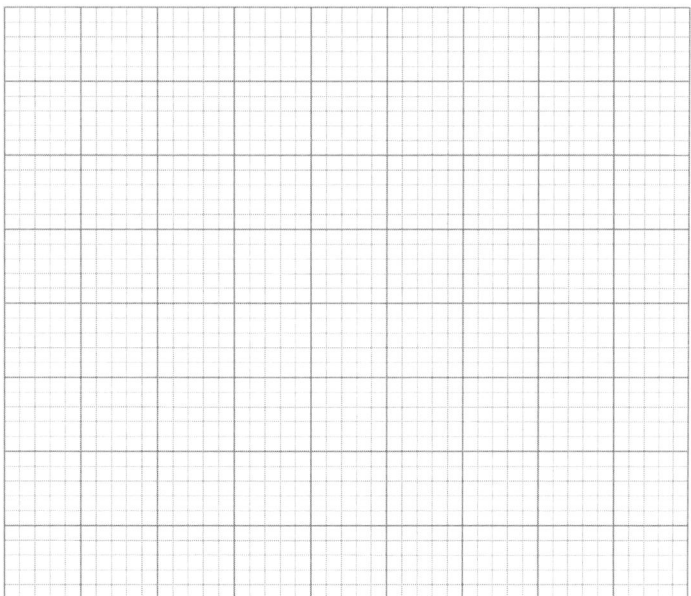

[5]

    b. Describe the shape of the graph of rate of reaction against temperature.

....

.... [2]

Link the words **A** to **D** on the left with descriptions **1** to **4** on the right.

A Anhydrous	**1** The water which makes up part of the crystal structure
B Hydrated	**2** Can go backwards or forwards
C Reversible	**3** Does not have water in it
D Water of crystallisation	**4** Has water in it

[2]

1. Hydrated and anhydrous cobalt(II) chloride can be converted to one another.

$$CoCl_2 \cdot 6H_2O \rightleftharpoons CoCl_2 + 6H_2O$$

pink (hydrated)   blue (anhydrous)

a. What is the meaning of the symbol $\rightleftharpoons$? .... [1]

b. How could you change pink cobalt(II) chloride to blue cobalt(II) chloride?

.... [1]

c. Suggest why the reaction: blue cobalt(II) chloride to pink cobalt(II) chloride is exothermic?

....

.... [2]

2. When a mixture of hydrogen and iodine is heated in a closed system, an equilibrium mixture with hydrogen iodide is formed.

$$H_2(g) + I_2(g) \rightleftharpoons 2HI(g)$$

Increasing the concentration of hydrogen or increasing the temperature shifts the position of equilibrium of this reaction.

a. State the meaning of 'closed system' and give an example.

....

.... [2]

b. Give two other characteristics of a reaction at equilibrium.

....

.... [2]

c. Give the meaning of the term 'position of equilibrium'.

.... [1]

Complete these sentences about the effect of temperature on equilibrium using words from the list.

endothermic   favours   heat   reverse   shifts

If a reaction is exothermic in the forward direction, it will be .... in the reverse direction. For an

exothermic reaction, when temperature increases, the equilibrium .... in the direction of

the .... reaction. It .... the endothermic change where .... is

taken in.   [5]

1. Sulfur dioxide reacts with oxygen to form an equilibrium mixture with sulfur trioxide.

$2SO_2(g)$ + $O_2(g)$ $\rightleftharpoons$ $2SO_3(g)$ $\Delta H = -197$ kJ / mol

Complete the following sentences about this reaction.

a. When oxygen is removed the position of equilibrium shifts to the ....   [1]

b. Decreasing the pressure shifts the position of equilibrium to the .... because there

   are .... moles of gas molecules in the equation on the ....   [3]

c. Decreasing the temperature shifts the equilibrium to the........................   [1]

2. In which direction does the position of equilibrium shift when the pressure on each of these reactions is increased?

a. $CO(g) + 2H_2(g) \rightleftharpoons CH_3OH(g)$ .... [1]

b. $CaCO_3(s) \rightleftharpoons CaO(s) + CO_2(g)$ .... [1]

c. $4HCl(g) + O_2(g) \rightleftharpoons 2H_2O(g) + 2Cl_2(g)$ .... [1]

d. $2HCl(g) \rightleftharpoons H_2(g) + Cl_2(g)$ .... [1]

3. When bismuth trichloride, $BiCl_3$, is added to water the following reaction occurs.

$BiCl_3(aq)$ + $H_2O(l)$ $\rightleftharpoons$ $BiClO(s)$ + $2HCl(aq)$

colourless solution   white precipitate

a. What would you see when concentrated HCl is added to the mixture?

   .... [1]

b. What would you see when a large volume of water was added to the reaction mixture?

   .... [1]

Link the words **A** to **D** on the left with descriptions **1** to **4** on the right.

A Oxidation	1 The removal of oxygen from a compound
B Reduction	2 The simultaneous oxidation and reduction in a chemical reaction
C Redox	3 The addition of oxygen to an element or compound
D Hydrogenation	4 An example of a reduction reaction

[2]

1. Draw arrows to show which of the elements or compounds have undergone oxidation and which have undergone reduction. An example is given below.

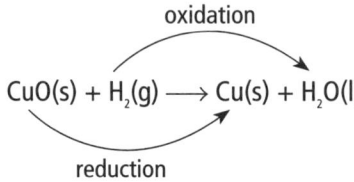

$$CuO(s) + H_2(g) \longrightarrow Cu(s) + H_2O(l)$$

oxidation

reduction

a. $2H_2(g) + O_2(g) \rightarrow 2H_2O(l)$ [2]

b. $PbO(s) + H_2(g) \rightarrow Pb(s) + H_2O(l)$ [2]

c. $Fe_2O_3(s) + 3C(g) \rightarrow 2Fe(s) + 3CO(g)$ [2]

d. $C(s) + H_2O(g) \rightarrow CO + H_2(g)$ [2]

e. $ZnO(s) + C(s) \rightarrow CO(g) + Zn(s)$ [2]

f. $3Fe(s) + 4H_2O(g) \rightarrow Fe_3O_4(s) + 4H_2(g)$ [2]

Match the beginning of each sentence, **A** to **D** to the endings **1** to **4**.

A The oxidation number of an uncombined element	**1** .... is the same as the charge on the ion.
B When a negative ion with a −1 charge is formed from an atom, the electron	**2** .... is zero.
C When an Fe^{2+} ion is converted to a Fe^{3+} ion an electron	**3** .... is added to the left of the half-equation.
D The oxidation number of a simple ion	**4** .... is removed from the iron(II) ion.

[2]

1. Identify the atom or ion which is reduced and the atom or ion which is oxidised in each of these equations.

   a. $2Mg + O_2 \rightarrow 2MgO$ [1]

   b. $PbO + H_2 \rightarrow Pb + H_2O$ [1]

   c. $2I^- + Cl_2 \rightarrow I_2 + 2Cl^-$ [1]

   d. $H_2O_2 + 2I^- + H^+ \rightarrow I_2 + 2H_2O$ [1]

2. Complete these half equations. Balance the charges by adding electrons.
   State whether oxidation or reduction has taken place.

   a. $Ca \rightarrow Ca^{2+} +$ .... Oxidation or reduction? .... [2]

   b. $Cl_2 +$ .... $\rightarrow$ .... Cl^- Oxidation or reduction? .... [2]

   c. $Al^{3+} +$ .... $\rightarrow$ .... Oxidation or reduction? .... [2]

   d. $Fe^{2+} \rightarrow Fe^{3+} +$ .... Oxidation or reduction? .... [2]

   e. $O_2 +$ .... $\rightarrow$ .... O^{2-} Oxidation or reduction? .... [2]

   f. $Pb^{4+} + 2e^- \rightarrow$ .... Oxidation or reduction? .... [2]

   g. .... $Br^- \rightarrow$ .... + ... Oxidation or reduction? .... [2]

3. Deduce the oxidation number of the underlined species.

   a. Na<u>Cl</u> .... 　　b. <u>N</u>H_3 .... 　　c. H_2<u>S</u>O_4 ....

   d. <u>Fe</u>_2O_3 .... 　　e. <u>P</u>_2O_5 .... 　　f. Na_2<u>S</u> .... [6]

Link the words **A** to **D** on the left with descriptions **1** to **4** on the right.

A Oxidation	**1** This happens when an atom gains one or more electrons
B Oxidising agent	**2** A substance which gives electrons to another substance
C Reducing agent	**3** This happens when an atom loses one or more electrons
D Reduction	**4** A substance which removes electrons from another substance [2]

1. Identify the oxidising and reducing agents in these equations.

   Underline the reducing agent and put a ring around the oxidising agent.

   a. $2Mg + O_2 \rightarrow 2MgO$ [1]

   b. $PbO + H_2 \rightarrow Pb + H_2O$ [1]

   c. $2I^- + Cl_2 \rightarrow I_2 + 2Cl^-$ [1]

   d. $H_2O_2 + 2I^- \rightarrow I_2 + 2H_2O$ [1]

2. **a.** Describe the colour change when excess potassium manganate(VII) is added to colourless sodium sulfite, Na_2SO_3, to produce colourless sodium sulfate, Na_2SO_4 .

   From .... to .... [1]

   b. Identify the oxidising agent in this reaction and explain your answer using changes in oxidation number.

   Oxidising agent   .... [1]

   Explanation   .... [2]

3. **a.** Describe how aqueous potassium iodide is used to reduce $Fe^{3+}(aq)$ ions to $Fe^{2+}(aq)$ ions.

   Test   ....

   Observations   .... [3]

   b. Name and give the formula of the coloured substance produced by oxidation of the iodide ions in this reaction.

   Name .... formula.. [2]

Language lab

Use the clues below to do this crossword

Across: 1. Element used to make a common acid

2. A pH above 7 is ....

5. An indicator used to test if a solution is acidic or alkaline

7. A substance which changes colour at a particular pH

Down: 1. A slightly alkaline substance we use to clean ourselves

2. A substance with a pH below 7

3. A substance with pH 7 is ....

4. Uni.........al Indicator is used to find the pH of a solution

6. A pH value less than 2   [9]

1. Link the phrases **A** to **E** on the left with the pH values **1** to **5** on the right.

| A   very acidic | 1   pH 6 |

| B   neutral | 2   pH 8 |

| C   very alkaline | 3   pH 1 |

| D   weakly acidic | 4   pH 14 |

| E   weakly alkaline | 5   pH 7 |

[3]

2. a. Link these acids and alkalis by choosing the correct formula from the list.

$Ca(OH)_2$ CH_3COOH H_2CO_3 HNO_3 H_3PO_4 H_2SO_4 $NaOH$ NH_3

i.   ammonia .... [1]   ii.   ethanoic acid .... [1]

iii. carbonic acid .... [1]   iv.   sulfuric acid .... [1]

v.   calcium hydroxide .... [1]   vi.   phosphoric acid .... [1]

vii. nitric acid .... [1]   viii. sodium hydroxide .... [1]

b. Name the indicator used to deduce the pH of a solution.

.... [1]

3. Many acids and alkalis are corrosive. What dos the word *corrosive* mean?

.... [1]

Language lab

Complete these sentences about acids using words from the list.

atoms   dioxide   hydrogen   hydroxides   replaced   salt   water

Acids react with many metals to form a salt and .... A salt is a substance formed

when particular hydrogen .... in an acid are .... by a metal. Acids also

react with some metal .... to produce a .... and water, and with carbonates

to produce a salt, ... and carbon ....   [7]

1. Link the colours of these indicators to the following pH values.

                    pH 3   pH 7   pH 13

   a   Litmus is blue at pH ....   [1]

   b. Universal indicator is green at pH ....   [1]

   c. Litmus is red at pH ....   [1]

2. Complete these word equations

   a. zinc oxide + hydrochloric acid → .... [1]

   b. iron + sulfuric acid → .... [1]

   c. sulfuric acid + lead carbonate → .... [1]

   d. hydrochloric acid + tin oxide → .... [1]

3. Complete the balanced chemical equations for these reactions.

   a. $Zn + H_2SO_4 →$ .... [1]

   b. $MgO + ...HNO_3 →$ .... [2]

   c. $CuCO_3 + ...HCl →$ .... [2]

   d. $Na_2CO_3 + ...HCl →$ .... [2]

   e. $Ca + ...HCl →$ .... [2]

   f. $......KOH + H_2SO_4 →$ .... [2]

4. Give the name and formula of the ion formed when an acid dissolves in water.

   Name .... Formula ....   [2]

Complete these sentences about bases using words from the list.

acid   alkalis   oxides   salt   soluble   sulfate   sulfuric

A base is a substance which reacts with an .... to form a salt. Bases which are ....

in water are called .... Many metal .... are basic because they react with

acids to form a .... and water. Ammonia reacts with .... acid to form

ammonium ...., so ammonia is also a base.   [7]

1. Complete these word equations

   a. sodium hydroxide + nitric acid → .... [1]

   b. calcium oxide + hydrochloric acid → .... [1]

   c. barium hydroxide + nitric acid → .... [1]

2. Complete the balanced chemical equations for these reactions.

   a. $MgO +$ ....$HNO_3 →$ .... [2]

   b. ....$NaOH + H_2SO_4 →$ .... [2]

   d. $Ca(OH)_2 +$ ....$HNO_3 →$ .... [2]

   e. ....$NH_3 + H_2SO_4 →$ .... [2]

3. Complete the equation below for the reaction of an alkali with an ammonium salt.

   $$Ca(OH)_2 + ...NH_4Cl → CaCl_2 + ...H_2O + ...$$ [2]

4. Explain why farmers add calcium oxide or calcium carbonate to the soil.

   ....

   ....

   .... [2]

5. Ammonia reacts with water: $NH_3(g) + H_2O(l) \rightleftharpoons NH_4OH(aq) + OH^-(aq)$

   How does this equation show that aqueous ammonia is an alkali?

   .... [1]

6. a. Give the name and formula of the ion present in all alkalis.

      Name .... Formula ....   [2]

   b. Complete the equation to represent neutralisation.

      $$H^+ + ... → ....$$ [2]

61

Language lab

Complete these sentences about strong and weak acids using words from the list.

dissociate   hydrogen   ions   molecules   partially   water

Aqueous solutions of acids contain .... ions. In strong acids the acid ....

ionise (........................) completely to form hydrogen .... and anions. When weak

acids dissolve in .... they become .... dissociated.   [6]

1. Define acids and bases in terms of proton transfer.

    An acid is a .... A base is a ....   [2]

2. Use the information in the table to answer parts **a**. and **b**.

acid 0.1 mol / dm³ solution	relative electrical conductivity	pH	rate of reaction with magnesium
ethanoic, CH_3COOH	0.5		
hydrochloric, HCl	25		
methanoic, HCOOH	2	2.4	
sulfuric, H_2SO_4	40		

   a. Hydrogen ions conduct electricity better than other ions. The greater the concentration of hydrogen ions, the better the electrical conductivity.

   Complete the third column of the table using the following pH values: pH 0.7, pH 1.0, pH 2.9.   [3]

   b. The rate of reaction of magnesium with different acids depends on the hydrogen ion concentration in the acid. Suggest whether the reaction of magnesium with each of these acids is fast or slow. Write your answers in the fourth column.   [4]

3. The equation below shows the ions present in the reactants and products of a neutralisation reaction.

    $H^+(aq) + NO_3^-(aq) + Na^+(aq) + OH^-(aq) \rightarrow NO_3^-(aq) + Na^+(aq) + H_2O(l)$

   a. Cancel out the spectator ions in this equation.   [1]

   b. Write the ionic equation for this reaction.

   .... [1]

   c. Explain, in terms of ions, why this is a neutralisation reaction.

   .... [1]

Link the words **A** to **D** on the left with the phrases **1** to **4** on the right.

A	End point

B	Indicator

C	Titre

D	Volumetric pipette

1	Used to deliver an accurate volume of liquid

2	The volume of acid delivered from a burette at the end-point of a titration

3	Substance which changes colour at the end point of an acid–base titration

4	When the indicator changes colour after adding a certain amount of acid

[2]

1. State the colour of each of these indicators at the pH values shown.

   a. litmus at pH 3 .... litmus at pH 8 .... [2]

   b. methyl orange at pH 3 .... methyl orange at pH 8 .... [2]

   c. thymolphthalein at pH 3 .... thymolphthalein at pH 12 .... [2]

2. The graph shows how the pH changes when an alkali is titrated with an acid.

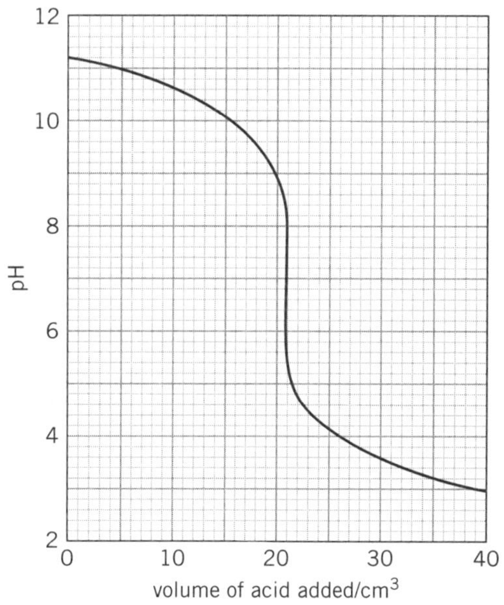

   a. State the value of the pH of the alkali at the beginning of the titration. .... [1]

   b. Deduce the volume of acid added when the pH was 7. .... [1]

   c. Describe exactly how the pH changes as the acid is added.

   ....

   ....

   .... [3]

63

Link the oxides **A** to **C** on the left with the descriptions **1** to **3** on the right.

A Acidic

B Amphoteric

C Basic

1 Oxides which react with acids but not with alkalis

2 Oxides which react with alkalis but not with acids

3 Oxides which react with both acids and alkalis

[2]

1. Complete these equations to show the reactions of some oxides with either acids or bases.

   a. $MgO +$ ....$HCl \rightarrow$ .... $+$ .... [2]

   b. $SO_2 +$ ....$NaOH \rightarrow Na_2SO_3 +$ .... [2]

   c. $CuO + H_2SO_4 \rightarrow$ .... $+$ .... [1]

   d. $CO_2 +$ .... $NaOH \rightarrow Na_2CO_3 +$ .... [2]

   e. $ZnO +$ ....$HNO_3 \rightarrow$ .... $+$ .... [2]

   f. $CaO + H_2SO_4 \rightarrow$ .... $+$ .... [1]

2. Complete these equations to show the reactions of some oxides with water to form acids or alkalis.

   a. $SO_2 + H_2O \rightarrow$ .... [1]

   b. $CO_2 + H_2O \rightarrow$ .... [1]

   c. $CaO + H_2O \rightarrow$ .... [1]

   d. $P_4O_6 +$ ....$H_2O \rightarrow$ .... H_3PO_3 [2]

   e. $Na_2O + H_2O \rightarrow$ .... [2]

3. Complete the equation for the reaction of zinc oxide with potassium hydroxide.

   $ZnO + 2KOH \rightarrow$ .... $+$ .... [2]

Language lab

Search for seven words involved in making a salt from a metal oxide. Names of pieces of apparatus are included. Words go forwards or downwards only. Write the words in the space below.

C	R	Y	S	T	A	L	X	A	S
E	N	C	R	I	P	Z	I	O	N
F	I	L	T	R	A	T	E	X	P
O	N	P	R	E	S	T	A	D	I
H	S	E	Z	F	I	L	T	E	R
E	O	X	I	D	E	Y	I	P	X
A	L	L	D	S	L	Y	M	O	L
T	U	P	T	P	P	R	E	D	P
X	B	I	M	Y	L	L	O	D	L
T	L	U	R	E	W	H	E	L	K
K	E	V	A	P	O	R	A	T	E

[7]

1. Zinc sulfate can be made by first warming sulfuric acid with excess zinc.

   a. How is a solution of zinc sulfate obtained from the reaction mixture?

   .... [1]

   b. The solution of zinc sulfate is crystallised. Describe how you could obtain pure dry crystals of zinc sulfate from a mixture of the crystals and remaining solution.

   ....

   ....

   .... [3]

2. Crystals of copper(*II*) sulfate can be made by warming excess copper(II) carbonate with sulfuric acid. Put the following stages in the correct order.

   A Allow the solution to cool and deposit crystals.

   B Pour the filtrate into an evaporating basin.

   C Wash and dry the crystals.

   D Filter the mixture to remove excess copper(*II*) oxide.

   E Warm the filtrate until the solution is very concentrated.

   F Filter off the crystals

   The order is .... [2]

3. Write symbol equations for the formation of these salts using suitable acids.

   a. Calcium sulfate from calcium oxide.

   .... [2]

   b. Zinc chloride from zinc.

   .... [2]

65

Use the clues below to do this crossword

Across: 1. The amount added from the burette in a titration

3. The solid dissolved to form a solution

6. This changes colour at the neutralisation point of a titration

7. This neutralises an alkali

8. 36 g of NaOH (Mr 40) is 0.X moles NaOH

9. The ....-point of a titration is where the indicator changes colour.

Down: 2. Procedure for finding the concentration of an alkali by adding acid from a burette

4. Long glass tube used in titrations

5. A volumetric ....is used to deliver an exact volume of solution [9]

1. The diagram shows the stages in making a soluble salt (sodium chloride) by neutralising an alkali (sodium hydroxide) with an acid (hydrochloric acid). The first three stages are shown in the diagram.

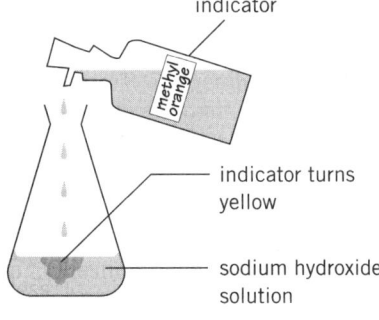

indicator

methyl orange

indicator turns yellow

sodium hydroxide solution

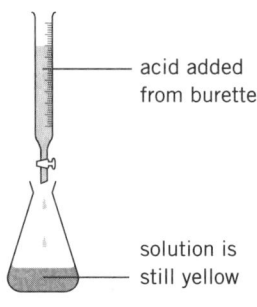

acid added from burette

solution is still yellow

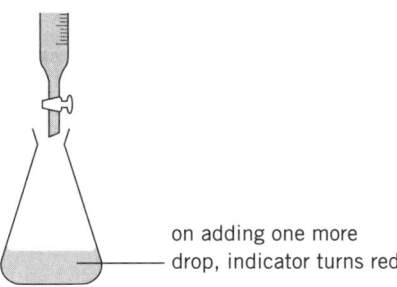

on adding one more drop, indicator turns red

Describe the next stages of the procedure to get colourless crystals of pure, dry sodium chloride crystals.

....

....

.... [5]

2. Write symbol equations for the formation of these salts using suitable acids.

a. Sodium nitrate from sodium hydroxide.

.... [2]

b. Ammonium sulfate from ammonia.

.... [2]

Complete these sentences about solubility using words from the list.

ammonium   carbonates   compounds   hydroxides   nitrates   precipitate   solutions

Salts such as ...., sodium salts and .... salts are soluble in water.

Many .... and .... are insoluble except those from Group I. An

insoluble substance formed when two .... of soluble .... are mixed

is called a ....   [7]

1. Are the following compounds soluble or insoluble? Write 'soluble' or 'insoluble' in the spaces provided. Use the solubility rules to help you.

   a. sodium hydroxide ....   [1]   b. potassium nitrate ....   [1]

   c. ammonium chloride ....   [1]   d. lead chloride ....   [1]

   e. iron(II) hydroxide ....   [1]   f. barium sulfate ....   [1]

2. Crystals of lead iodide can be made from solutions of lead nitrate and potassium iodide. Put the following stages in the correct order.

   A Filter the mixture.

   B Make up aqueous solutions of lead nitrate and potassium iodide

   C Dry the residue of lead iodide in a warm oven.

   D Rinse the residue on the filter paper with distilled water.

   E Mix the solutions. A yellow precipitate forms.

   The order is ....   [2]

3. a. The equation below shows the ions present in the reactants and products of a precipitation reaction.

   $Ag^+(aq) + NO_3^-(aq) + K^+(aq) + Br^-(aq) \rightarrow AgBr(s) + NO_3^-(aq) + K^+(aq)$

   i. Cancel out the spectator ions in this equation.   [1]

   ii. Write the ionic equation for this reaction.

   ....   [1]

   b. Write an ionic equation for this reaction. Include state symbols.

   $BaCl_2(aq) + K_2SO_4(aq) \rightarrow BaSO_4(s) + 2KCl(aq)$

   ....   [2]

Language lab

Complete these sentences about collecting gases using words from the list.

air   collect   displacement   downward   less   upward

When a gas is denser than ...., you collect it by .... displacement of air.

If a gas is .... dense than air, you .... it by ....

.... of air.   [6]

1. The diagram shows four ways of collecting gases in the laboratory.

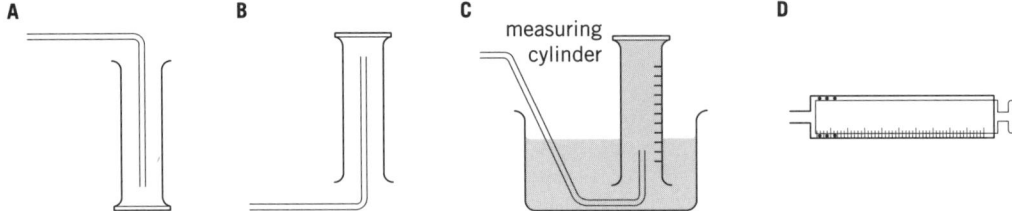

A   B   C   measuring cylinder   D

   a. Which method of gas collection, **A**, **B**, **C** or **D** is used for:

      i.   Measuring the volume of a gas which is sparingly soluble in water .... [1]

      ii. Collecting a gas which is lighter than air .... [1]

      iii. Collecting a gas which is heavier than air? .... [1]

   b. Which method of gas collection, **A**, **B**, **C** or **D** involves:

      i.   Upward displacement of air, ... [1]

      ii. Collection in a gas syringe? .... [1]

2. Link the gases **A** to **D** on the left with the best test results **1** to **4** on the right.

A Ammonia	**1** Bleaches damp litmus paper
B Carbon dioxide	**2** Turns damp red litmus paper blue
C Chlorine	**3** Turns acidified aqueous potassium manganate(VII) colourless
D Sulfur dioxide	**4** Turns limewater milky

[2]

Language lab

Complete these sentences about flame tests using words from the list.

coloured	edge	flame	lithium	luminous	potassium	sample	wire

A .... is put on the end of a platinum .... and placed at the .... of a

non-........................ Bunsen .... If .... is present, the flame is

.... red. If .... is present, the flame has a lilac colour.   [8]

1. Match the flame colours in the list with the ions **a** to **d**.

blue-green   orange-red   pale green   yellow

a. barium ....   **b.** calcium ....

c. copper ....   **d.** sodium ....   [4]

2. Complete the table showing the observations when aqueous solutions of ions react with aqueous sodium hydroxide or aqueous ammonia.   [16]

aqueous ion	reaction with sodium hydroxide	reaction with aqueous ammonia
$Al^{3+}(aq)$	At first .... In excess	At first .... In excess ....
$Cr^{3+}(aq)$	At first .... In excess	At first .... In excess ....
$Cu^{2+}(aq)$	At first .... In excess	At first .... In excess ....
$Fe^{3+}(aq)$	At first .... In excess	At first .... In excess ....

Language lab

Complete these sentences about the test for halides using words from the list.

chloride   cream   drops   nitrate   precipitate   solution

A few .... of nitric acid are added to the .... thought to be a halide.

Aqueous silver .... is then added. If a .... is present, a white .... is

seen. If a bromide is present, a ....-coloured precipitate is seen.   [6]

1. Link the anions **A** to **D** on the left with the test results **1** to **4** on the right.

A Carbonate ions	**1** Ammonia produced when heated with sodium hydroxide and aluminium foil
B Aqueous nitrate ions	**2** White precipitate formed on addition of aqueous nitric acid and barium nitrate
C Aqueous sulfite ions	**3** SO_2 produced when warmed with dilute hydrochloric acid
D Aqueous sulfate ions	**4** Effervescence of carbon dioxide on the addition of an acid

[2]

2. **a.** Write a symbol equation for the reaction of aqueous silver nitrate with aqueous sodium chloride. Include state symbols.

.... [2]

b. Convert this equation into an ionic equation.

.... [1]

3. **a.** Write a symbol equation for the reaction of aqueous barium chloride with aqueous sodium sulfate. Include state symbols.

.... [3]

b. Convert this equation into an ionic equation.

.... [1]

Use the clues below to do this crossword

Across: 4. Symbol for metal in Group II which is more reactive than calcium

5. Green gas

6. Element in Group III having lowest A_r

8. A transition element with fewer electrons than copper

Down: 1. Metal in Group II and Period 3

2. Symbol for argon

3. Its ions give yellow colour to a flame

4. Yellow element in Group VI

5. This element is found as graphite

7. Symbol for element which has an amphoteric oxide

[10]

1. The table shows some information about some of the elements in Period 3.

element	Na	Mg	Al	Si	P	S	Cl
electronic configuration	2,8,1						
melting point / °C	98	649	660	1410	590	119	−101
formula of typical compounds	NaCl Na$_2$O	MgCl$_2$ MgO	AlCl$_3$	SiCl$_4$ SiH$_4$	PCl$_3$ PH$_3$	H$_2$S	HCl

a. Complete the 2nd and 4th lines of the table. [4]

b. i. Describe how the melting points of the elements change across the Period.

.... [2]

ii. What type of structures are Na, Mg and Al?

.... [2]

iii. Explain in terms of structure and bonding why Si has the highest melting point in this Period.

....

.... [2]

iv. Explain in terms of structure and bonding why the melting points P, S and Cl are relatively low.

....

.... [2]

Language lab

Complete the sentences about Group I elements using words from the list.

alkali   densities   hydrogen   hydroxide   melting

The .... metals have low .... points and .... compared with most other metals.

They react with water to form a metal .... and ....   [5]

1. The table shows some properties of some Group I metals.

group I metal	density in g / cm³	melting point / °C	metallic radius / nm	observations when the metal reacts with water
lithium	0.53	181	0.157	Moves over the surface very slowly Fizzes gently Does not melt or go into a ball Does not burst into flame
sodium	0.97	98	0.191	....
potassium	0.86	....	0.235	Moves over surface very rapidly Fizzes very rapidly Melts and goes into a ball then bursts into flame Slight 'pop' when reaction near the end
rubidium	1.53	39	....	....

a. Complete the table.   [8]

b. Caesium is below rubidium in the Periodic Table. Predict a value for the density of caesium.

    .... [1]

Complete these sentences about the displacement reactions of halogens using words from the list.

bromine   colourless   halide   halogen   less   more   orange

When aqueous chlorine is added to a .... solution of potassium bromide, the solution turns ....

because .... has been displaced. This is because a .... reactive .... displaces a

.... reactive halogen from an aqueous solution of its ....   [7]

1. The table shows some properties of fluorine, chlorine, bromine and iodine.

halogen	melting point / °C	boiling point / °C	state at −40 °C	depth of colour	atomic radius / nm
fluorine	−220	−188			
chlorine	−101	−35			
bromine	−7	59			
iodine	114	184			

a. Describe the trend in the melting points of the halogens.

.... [1]

b. Use the values of the melting and boiling points in the table to deduce the state of the halogens at −40 °C. Write your answers in the table.   [4]

c. Draw an arrow in the 5th column to show the trend in the depth of colour (light → dark)   [1]

d. Draw an arrow in the 6th column to show the trend in atomic radius (smaller → larger)   [1]

2. State the observations when an aqueous solution of bromine is added to an aqueous solution of potassium iodide. Explain these observations.

....

....

.... [4]

Complete these sentences about the noble gases using words from the list.

atoms   eight   electron   gaining   more   noble   sharing   shell   stable

The Group VIII gases (............... gases) are unreactive because their .... configurations make them

.... It is difficult for their .... to form ionic bonds by .... or losing electrons or covalent

bonds by .... electrons. Helium is particularly unreactive because there cannot be .... than two

electrons in the first .... The other gases have .... electrons in their outer shell which is a stable

electronic configuration.   [9]

1. Use ideas about electronic configuration to suggest why the Group VIII elements are monatomic and not diatomic.

   ....

   .... [2]

2. The table shows the melting points and electrical conductivity of the Period 3 elements.

element	Na	Mg	Al	Si	P	S	Cl
melting point / °C	98	649	660	1410	590	119	−101
conductivity in S/m	0.218	0.224	0.382	very low	very low	very low	very low

   a. Describe the change in melting point across Period 3.

      .... [2]

   b. Explain the change in melting point in terms of structure and bonding.

      ....

      ....

      ....

      .... [4]

   c. Explain why Si, P, S and Cl are poor electrical conductors but Na, Mg and Al are good conductors.

      ....

      .... [2]

   d. Suggest why electrical conductivity increases from Na to Al.

      .... [2]

Use the clues below to do this crossword

Across: 2.   First three letters of atom with 27 electrons

4.   A property of transition element ions

6.   Abbreviation for oxidation number?

8.   Iron, cobalt and nickel are magn ....

9.   Symbol for nickel

10.   Property of transition element compound which speeds up a reaction

Down: 1.   Metal used for making bridges and cars

2.   .... plating on bicycles and taps

3.   Iron is hard, sodium is ....

4.   Ions of this element give a dark blue solution with excess aqueous ammonia

5.   Transition elements have a very high ....

7.   Copper has a high melting ....

[12]

1. The boxes below show some properties of a non-transition element and of a transition element. The boxes are muddled up. (M = metal)

| A   Melting point 1890 °C | B   Forms a chloride of formula MCl_2 only |

| C   Density 3.51 g / cm³ | D   Forms chlorides which are pink and green |

| E   Forms chloride of type MCl_2, MCl_3 and MCl_4 | F   Forms a colourless chloride |

| G   A compound of M is a good catalyst | H   Density 5.96 g / cm³ |

| I   Melting point 725 °C | J   Compounds of M show no catalytic activity |

a. Which letters represent transition elements?

.... [3]

b. Give two other typical properties of transition elements which are not mentioned in part **a**.

.... [2]

2. Write the formulae of the transition element ions in the following compounds.

a. Ag_2O ....   b. $CuSO_4$ .... [2]

c. $Cr(NO_3)_3$ ....   d. $Fe_2(SO_4)_3$ .... [2]

3. Write an ionic equation, including state symbols, for the reaction of aqueous iron(III) ions with aqueous hydroxide ions.

.... [3]

Complete these sentences about the reactivity of metals using words from the list.

alkaline   blue   cold   electrons   hydrogen   lower   oxide

The products formed by metals which react with .... water are a metal hydroxide and ....

The hydroxides are .... and so turn red litmus .... The products formed by metals which

only react with steam are a metal .... and hydrogen. Copper does not react with water because it is

.... in the reactivity series than hydrogen and cannot take the .... away from the

hydrogen in the water.   [7]

1. Complete these equations for the reactions of metals with water or steam.

    a. ....$Na(s)$ + ...$H_2O(l)$ $\rightarrow$ .... + ...   [3]

    b. ....$Fe(s)$ + ...$H_2O(g)$. $\rightarrow$ ....$Fe_3O_4(s)$ + ...   [3]

2. Some observations for the reaction of metals with water are given in the table.

metal	observations
barium	
calcium	Gives off bubbles rapidly with cold water, disappears quite quickly
lead	
magnesium	Gives a few bubbles with hot water, disappears slowly
zinc	Only reacts when heated to red-hot with steam

    a. Put barium, calcium and magnesium and zinc in order of their reactivity. Put the most reactive first.

        .... [1]

    b. Barium is more reactive than calcium and lead is less reactive than zinc.

        Write the observations for barium and lead in the table above.   [3]

3. Put these metals in order of their reactivity using the information below.

metal	concentration of HCl(aq) in mol/ dm³	observations
iron	0.5	slow bubbling
lead	6.0	slow bubbling
lithium	0.5	rapid stream of bubbles
magnesium	0.5	steady stream of bubbles

    least reactive .... most reactive   [2]

Complete these sentences about the reactivity of aluminium using words from the list.

acids   freshly   layer   oxide   oxygen   reactivity   sticks   unreactive

Although aluminium is high in the .... series, samples of the metal exposed to the air do not appear to

react with water or dilute .... This is because .... -made aluminium reacts with

.... in the air to form a thin .... of aluminium .... on its surface, which is

relatively .... The oxide layer .... to the metal surface strongly so is not easily removed. [8]

1. Identify the reducing agent and the oxidising agent in each of these equations.

a. Fe(s) + CuO(s) → FeO(s) + Cu(s)

reducing agent .... oxidising agent ....   [1]

b. Fe_2O_3(s) + 3Mg(s)   →   2Fe(s) + 3MgO(s)

reducing agent .... oxidising agent ....   [1]

2. Manganese is extracted by reduction of manganese oxide with hot aluminium.

Write a word equation for this reaction.

.... [1]

3. Complete these equations for the reduction of some metal oxides to metals.

a. SnO_2 + ...C → .... + ...   [2]

b. ....NiO + CO + H_2 → .... + ... + H_2O [2]

c. ....PbO + C → .... + ...   [2]

4. The table shows the ease of reduction of some metal oxides.

metal oxide	ease of reduction
chromium(III) oxide	reduced by carbon at 1200 °C
copper(II) oxide	reduced by carbon below 400 °C
manganese(II) oxide	reduced by carbon at 1400 °C
tin(IV) oxide	reduced by carbon at 400 °C
silver(I) oxide	reduced by heating alone at low temperatures

a. Put the metals in order of their reactivity.

least reactive .... most reactive   [2]

b. Construct a balanced equation for the reduction of chromium(III) oxide, Cr_2O_3, by carbon to form chromium and carbon monoxide.

.... [2]

77

Complete these sentences about the reactivity series using words from the list.

bauxite   bonds   carbon   electrolysis   extracted   hematite   reactive   temperature

Metals above carbon in the reactivity series are not usually .... from their oxides by heating with

.... This is because the metal .... to the oxygen too strongly. Carbon is not ....

enough to remove the oxygen from the metal oxide unless a very high .... is used. Iron which is below

carbon in the reactivity series is extracted from .... ore. Aluminium, which is above carbon in the

reactivity series is extracted by .... of aluminium oxide obtained from .... ore.   [8]

1. Part of the reactivity series is shown on the left but some of the metals are missing.

   a. Complete the series in the correct order of reactivity using the list of metals on the right.

potassium	aluminium
....	
calcium	copper
....	
....	gold
carbon	
zinc	iron
....	
hydrogen	magnesium
....	
silver	sodium
....	

   [3]

   b. Which of the metals in part **a** are extracted by electrolysis of their oxides and not by reduction with carbon? Explain your answer.

   ....

   .... [2]

2. Complete these equations for the reduction of some metal oxides to metals.

   a. Fe_2O_3 + ...CO → .... + ...CO_2 [2]

   b. ZnO + CO → .... + ...   [1]

   c. ....CuO + C → .... Cu + CO_2 [1]

3. Name the reducing agent in question **2.b**.

   .... [1]

Language lab

Complete these sentences about the reactivity of metals using words from the list.

          aqueous   atoms   displaces   higher   losing   solution

Calcium is .... in the reactivity series than copper. So calcium atoms are better at ....

electrons than copper. When calcium is added to .... copper(II) sulfate, calcium ....

the copper and copper metal is formed. Calcium .... are converted to calcium ions which go

into ....   [6]

1. The diagram shows different metals are placed in solutions of metal salts.

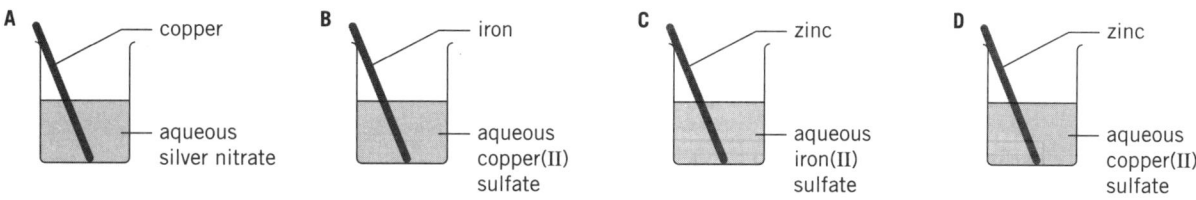

Some of the results are shown in the table.

experiment	colour at the start		colour after 20 minutes	
A	metal	brown	metal	silver-grey surface
	solution	colourless	solution	blue
B	metal	silvery grey	metal	
	solution		solution	
C	metal	grey	metal	
	solution	light green	solution	colourless
D	metal		metal	
	solution		solution	

a. Complete the table to show the colour changes.   [8]

b. Use the results to the metals in order of their reactivity.

   least reactive..most reactive   [1]

c. Explain why there is no colour change when a copper rod is placed in aqueous zinc sulfate.

   .... [1]

Language lab

Search for the names of eight words related to the extraction of iron. Words go forwards or downwards only. Write the names in the space below.

B	N	H	H	A	T	E	S	C
L	S	E	C	O	K	E	X	A
A	P	M	S	X	P	Q	Q	R
S	C	A	T	I	R	O	N	B
T	G	T	E	D	R	R	S	O
X	O	I	L	E	V	F	L	N
V	E	T	L	V	B	Y	A	L
N	L	E	E	O	R	I	G	N
R	E	D	U	C	T	I	O	N

[8]

1. The diagram below shows a blast furnace for the extraction of iron.

   On the diagram draw arrows and the following letters to show:

   A → where air is blown into the furnace

   B → where the iron ore is added to the furnace

   C → where the molten iron is removed

   D → where the slag is removed

   E → where waste gases exit the furnace.

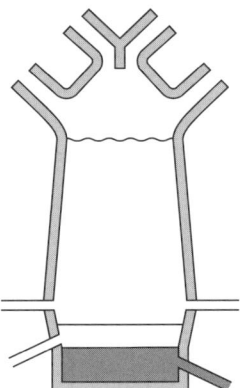

[5]

2. Iron(III) oxide is reduced in the furnace by carbon monoxide.

   Describe the two stages by which this carbon monoxide is produced.

   ....

   .... [2]

3. Complete these sentences about the use of limestone in the blast furnace using words from the list.

   floats   impurity   molten   oxide   silicate   silicon   slag   thermal

   Limestone undergoes .... decomposition to form calcium .... This reacts with

   .... dioxide (sand) which is an .... in the ore. The calcium ....

   (...................) formed runs down the furnace and .... on top of the .... iron.   [8]

4. Complete this equation for the reduction of iron oxide to iron.

   $Fe_2O_3(s)$ + ...CO → .... + ...   [2]

Complete these sentences about rusting using words from the list.

corrodes   electrons   ions   iron   more   rusting   sacrificial   solution

Blocks of zinc can be placed on the hull of a ship to stop it .... Zinc is ....

reactive than .... so it loses .... and forms .... more easily than iron.

The zinc ions go into .... and so the zinc .... instead of the iron. This is called

.... protection.   [8]

1. The graph shows the rate of corrosion of iron at different pH values in aerated water.

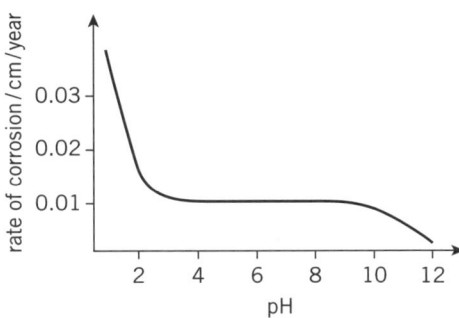

a. Describe how the corrosion of iron varies with pH.

....

.... [3]

b. At acidic pH values 'green rust' is formed. At less acidic pH values 'green rust' is converted to 'red rust'.

The simplified formula of 'red rust' is $Fe_2O_3 \cdot xH_2O$.

Give the name of the compound $Fe_2O_3 \cdot xH_2O$.. [1]

2. Suggest reasons for the following:

a. An iron object in the desert rusts very slowly.

.... [1]

b. Painting an iron object stops it from rusting.

.... [2]

c. Galvanising an iron object. State and explain two reasons.

....

....

.... [4]

Language lab

Link the metals **A** to **D** on the left with the phrases **1** to **4** on the right.

| A   Aluminium aircraft alloy |
| B   Brass |
| C   Chromium and nickel |
| D   Stainless steel |

| 1   A mixture of Cu and Zn |
| 2   A mixture of Fe, C and other elements |
| 3   These are found in stainless steel |
| 4   Contains Al, Cu and Mg |

[2]

1. Which one of the diagrams **A** to **D** best represents an alloy. .... [1]

A   B   C   D

2. Complete the sentences about alloys using words from the list.

alloyed   difference   force   harder   layers   mixtures   regular   prevents

Alloys are .... of metals or mixtures of metals with non-metals. Alloys are often

.... and stronger than pure metals. When a metal is .... with another metal, the

.... in the size of the metal atoms makes the arrangement of the .... in the lattice

less .... This .... the layers from sliding over each other as easily when a ....

is applied.   [8]

3. Complete the table about the uses and properties of different metals and alloys.

alloy	use	properties which makes it suitable for the use
aluminium alloy	aircraft body	1... [1] 2... [1]
stainless steel	cutlery	1... [1] 2... [1]

4. Suggest why an alloy of tin and lead has lower melting point than either pure tin or pure lead.

....

.... [2]

Link the start of sentences **A** to **D** on the left with the endings **1** to **4** on the right.

All the sentences are about the recycling of metals.

A   Recycling reduces the amount of dust ....

1   .... so there are fewer problems with disposal of unwanted materials.

B   Recycling reduces the amount of carbon dioxide produced ....

2   .... so that land can be used for other purposes such as agriculture.

C   Recycling reduces the amount of waste ....

3   .... from both mining and extracting the metal using carbon.

D   Recycling reduces the need for mining ores ....

4   .... caused by mining the ore.

[2]

1. Complete the table about the uses and properties of different metals. (For tungsten and brass, think about the use and suggest properties which might be related to the use.)

metal	use	properties which makes it suitable for the use
Aluminium	Food containers	1. .... [1] 2. .... [1]
Aluminium	Overhead electrical cables	1. .... [1] 2. .... [1]
Copper	Electrical wiring	1. .... [1] 2. .... [1]
Stainless steel	1. .... [1] 2. .... [1]	.... [1]
Tungsten steel	For drilling other metals	1. .... [1] 2. .... [1]
Brass	Door handles	1. .... [1] 2. .... [1]

Search for the names of seven words related to the manufacture of ammonia. Words go forwards or downwards only. Write the names in the space below.

C	A	T	A	L	Y	S	T	H
X	C	P	T	O	S	D	F	Y
Q	W	E	I	R	T	Y	U	D
N	I	T	R	O	G	E	N	R
A	H	D	O	F	G	H	J	O
R	A	W	N	C	V	B	M	G
F	B	M	E	T	H	A	N	E
V	E	E	D	C	T	Y	U	N
P	R	E	S	S	U	R	E	C

[7]

1. State the source of hydrogen used to make ammonia.

.... [1]

2. Complete the equation for the synthesis of ammonia.

$N_2(g)$ + ...(g) $\rightleftharpoons$ ....(g) [2]

3. The graph below shows the effect of temperature and pressure on the percentage yield of ammonia.

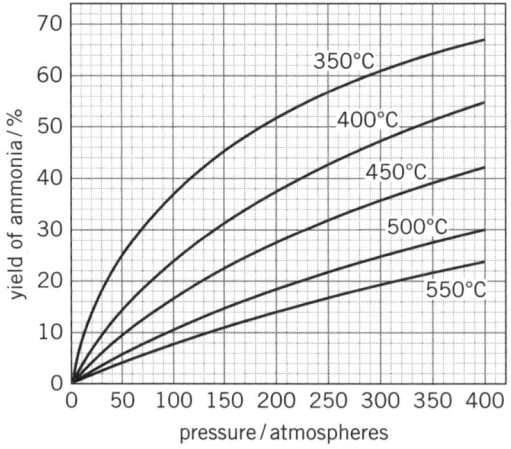

a. Describe the effect of pressure on the percentage yield.

.... [1]

b. How does the data in the graph show that the reaction is exothermic?

.... [1]

c. Deduce the percentage yield of ammonia at 200 atmospheres and 350 °C?

.... [1]

d. State one advantage and one disadvantage of using a high temperature in the reaction.

....

.... [2]

Language lab

Complete these sentences about fertilisers using words from the list.

fertilisers   nitrates   nutrients   phosphates   phosphorus   proteins   salts

For healthy growth crop plants need three major elements, nitrogen, ... and potassium.

Plants take up these elements in the form of nitrates, ... and potassium ....

The .... are needed to make .... for growth. Farmers add .... to

the soil to add back the .... which plants have absorbed for growth.   [7]

1. A flow chart for making fertilisers is shown below.

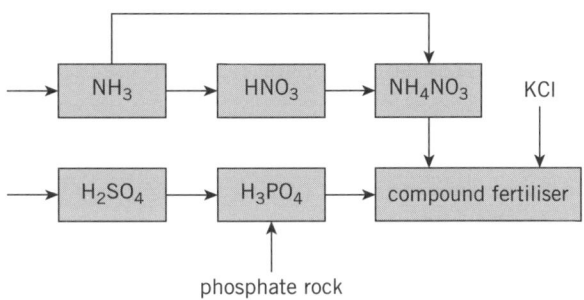

a. Name the compounds in the diagram:

NH_3 .... HNO_3 ....

NH_4NO_3 .... H_2SO_4 ....

H_3PO_4 .... KCl ....   [6]

b. Write a word equation for the reaction between NH_3 and HNO_3.

.... [1]

c. Name a suitable acid and base for making these fertilisers.

i. Ammonium sulfate .... [2]

ii. Potassium chloride .... [2]

iii. Sodium phosphate .... [2]

2. a. Farmers add lime to their fields. Suggest why they do this.

.... [1]

b. Lime reacts with water in the soil to form calcium hydroxide, $Ca(OH)_2$. Calcium hydroxide can react with ammonium sulfate fertiliser, $(NH_4)_2SO_4$, in the soil. Write a chemical equation for this reaction.

.... [2]

Search for the names of seven words related to the manufacture of sulfuric acid. Words go forwards or downwards only. Write the names in the space below.

A	B	C	O	N	T	A	C	T	Z
R	T	Y	C	U	R	S	H	R	R
V	A	N	A	D	I	U	M	X	R
O	Q	V	T	T	O	A	S	D	F
L	W	B	A	O	X	Y	G	E	N
E	E	N	L	Y	I	A	S	D	F
U	R	M	Y	H	D	D	F	G	H
M	T	S	S	J	E	H	J	K	G
E	X	O	T	H	E	R	M	I	C

[7]

1. In the Contact process, sulfur dioxide is converted to sulfur trioxide in the presence of vanadium(V) oxide.

$$2SO_2(g) \ + \ O_2(g) \ \rightleftharpoons \ 2SO_3(g)$$

   a. What is the purpose of the vanadium(V) oxide?

   .... [1]

   b. The graph below shows the percentage conversion of SO_2 to SO_3 at different temperatures. The pressure was just above atmospheric pressure.

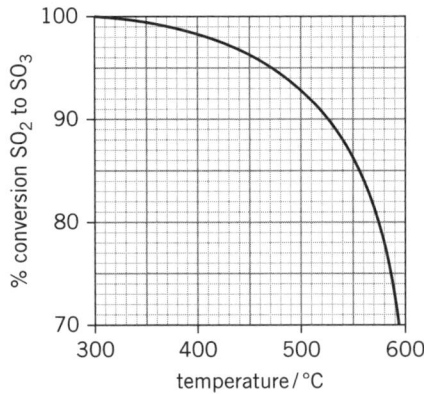

   i. Describe in detail the effect of temperature on the percentage conversion of SO_2 to SO_3.

   .... [2]

   ii. Deduce the percentage conversion of SO_2 to SO_3 at 500 °C.

   .... [1]

   iii. How does the data in the graph show that the reaction is exothermic?

   .... [1]

   c. Predict the effect of increasing the pressure on this reaction. Explain your answer.

   ....

   .... [2]

Complete these sentences about acid rain using words from the list.

| burnt | dioxide | lakes | leaves | oxidised | pH | sulfuric | sulfurous | vapour |

When fossil fuels containing sulfur are ...., sulfur .... gas is formed. Sulfur

dioxide reacts with water .... in the atmosphere to form .... acid. Some of

the sulfur dioxide is .... in the atmosphere to form sulfur trioxide. This reacts with water

vapour to produce .... acid. Acid rain has a .... of about 4.5 and so can

harm wildlife in .... and damage the .... of trees.   [9]

1. Oxides of nitrogen can cause acid rain.

   a. Name one source of oxides of nitrogen in the air.

   .... [1]

   b. Name an acid formed when oxides of nitrogen react with water vapour in the air.

   .... [1]

2. The graph below shows the concentration of nitrogen dioxide in the air in the streets of a large city throughout a particular day.

   Describe and explain the shape of this graph.

   ....

   ....

   ....

   .... [4]

3. Nitrogen dioxide is one of the oxides responsible for acid rain.

   Describe and explain the effect of acid rain on a building made of limestone.

   ....

   .... [3]

Language lab

Link the words **A** to **D** on the left with the phrases **1** to **4** on the right.

A Methane		1 Responsible for increase global warming
B Particulates		2 A smoky fog
C Photochemical		3 Tiny particles which can cause cancer
D Smog		4 Reaction which involves light

[2]

1. Match the percentages below to the percentage of the gases **a.** to **d.** in the atmosphere.

    0.04%   about 1%   21%   78%

    a. carbon dioxide ....%   **b.** nitrogen .... %

    c. noble gases ....%   **d.** oxygen .... %   [2]

2. Complete these sentences about the sources of carbon monoxide and sulfur dioxide in the atmosphere using words from the list below. (Not all the words are used.)

    burn   carbon   excess   fossil   gaseous   limited   oxygen   sulfur

    Carbon monoxide is formed when .... -containing compounds .... in a

    .... supply of air. Sulfur dioxide is formed when .... fuels containing

    .... burn in air.   [5]

3. Give one harmful effect of

    a. Increased concentration of carbon dioxide in the atmosphere.

    .... [1]

    b. Carbon monoxide on humans .... [1]

4. The table shows the percentage of some of the gases in dry air in 1985 and 2015

gas	volume in 100 cm³ air in 1985	volume in 100 cm³ air in 2015
nitrogen	78.082	78.081
oxygen	20.950	20.949
carbon dioxide	0.034	0.0397
methane	0.00014	0.000179

    a. Which two gases have shown the largest percentage change since 1985?

    .... and ....   [1]

    b. What other substance, which is liquid at r.t.p, is present in variable amounts in the air?

    .... [1]

Language lab

Use the clues below to do this crossword.

Across: 2. Photo......... produces most of the oxygen in the atmosphere

3. The release of energy from glucose in living things

6. A fossil .... when burnt releases CO_2

7. The number of C atoms in one molecule of glucose

8. A product of photosynthesis

Down: 1. Breakdown of vegetation by bacteria

2. Sea creatures such as shellfish have .... made out of calcium carbonate

4. Large seas which absorb huge amounts of carbon dioxide

5. A solid fossil fuel

[9]

1. The diagram shows the carbon cycle. The numbers are the relative amounts of carbon transferred per year.

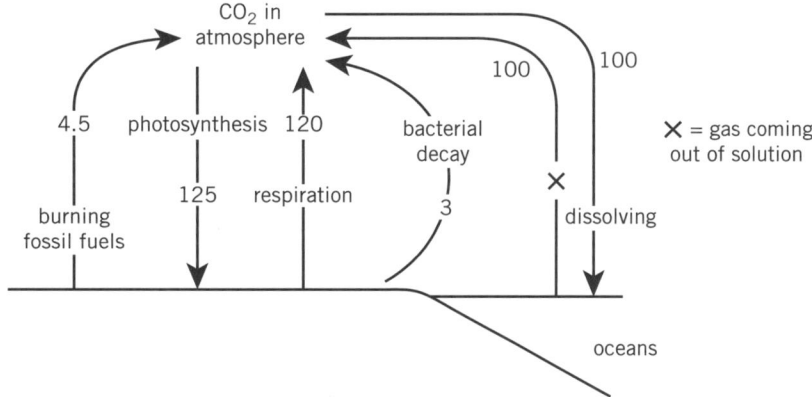

a. Comment on the balance between respiration and photosynthesis in this cycle.

....

.... [1]

b. Describe two things which might upset the balance of the carbon cycle.

....

.... [2]

2. Write a word equation and complete the symbol equation for photosynthesis.

word equation .... [2]

symbol equation .... + ... → $C_6H_{12}O_6$ + 6....... [2]

Language lab

Link the words **A** to **D** on the left to their meanings **1** to **4** on the right.

A Absorption		1 To do with heat
B Emission		2 Bouncing back
C Reflection		3 Given out
D Thermal		4 Taken in

[2]

1. The graphs show the concentration of carbon dioxide in the atmosphere and the estimated global mean (average) temperature over a period of 120 years.

   a. Carbon dioxide is a greenhouse gas. State the meaning of the term *greenhouse gas*.

   .... [2]

   b. How does the information from the graphs support the idea that carbon dioxide is a greenhouse gas?

   ....

   .... [2]

2. The absorption of energy by greenhouse gases leads to global warming.

   Give two effects of global warming.

   ....

   .... [2]

3. Complete these sentences about methane using words from the list.

   absorbs   atmosphere   bacterial   digestion   vegetation

   Methane is a greenhouse gas which is formed by the .... decomposition of .... and

   as a waste product of .... in animals. It is present in the .... at a lower concentration

   than carbon dioxide but it .... much more thermal energy per mole.   [5]

Link the words **A** to **D** on the left to their meanings **1** to **4** on the right.

A Desulfurisation	**1** Farm animals
B Flue gas	**2** Examples are solar power and wind power
C Livestock	**3** Removal of sulfur from a fuel
D Renewable energy	**4** SO_2 and other gases produced by burning fuels in furnaces

[2]

1. **a.** Explain the importance of a catalytic converter attached to a car exhaust.

    ....

    ....

    .... [3]

    b. Complete these equations for the reactions in a catalytic converter.

    i. ....NO_2 → .... + ...O_2 [2]

    ii. ....NO + ... CO → .... + ... [2]

2. Complete these sentences about flue gas desulfurisation using words from the list.

    carbonate   combustion   fuels   neutralise   sulfite   sulfur   waste

    Flue gas desulfurisation is the process of removing .... dioxide from the gases

    formed during the .... of fossil .... in furnaces. The ....

    gases are passed through moist calcium .... or calcium oxide. These compounds

    .... the acidic sulfur dioxide. Solid calcium .... is formed. [7]

3. **a.** State the meaning of the term *climate change* and explain why it is important to reduce its effects.

    ....

    .... [3]

    b. Explain why both planting more trees and using renewable energy can reduce the effects of climate change.

    i. Planting more trees ....

    .... [2]

    ii. Using renewable energy ....

    .... [2]

Complete these sentences about water treatment using words from the list.

bacteria   branches   filter   harmful   insoluble   plant   settle   smells

In a water treatment ...., large objects such as plant .... are first trapped by metal screens. Other

solid particles are then left to .... to the bottom of the tank. The water is then passed through a ....

made of sand or gravel. This removes small .... particles. Carbon is added to remove bad ....

Chlorine is then added to the water to kill .... which may be .... to health.   [8]

1. Water from natural sources contains oxygen and mineral salts. State the beneficial effects of these substances.

Oxygen ....

Mineral salts ....   [2]

2. The table shows the concentration in mg / dm³ of some ions present in water from three different sources.

ion	seawater	rain water	river water
Na^+	10000	9	11
Ca^{2+}	900	2	1
K^+	1000	1	4
SiO_3^{2-}	500	0.5	7
Cl^-	17000	16	12
HCO_3^-	700	3	2
NO_3^-	trace	0.01	3

a. Which positive ion in seawater in the table is present at the lowest concentration?

....   [1]

b. What are the major differences between rainwater and river water in terms of the concentration of the ions present?

....

....

....   [3]

c. Calculate the concentration of chloride ions in 200 cm³ of the river water.

Concentration = ... mg / dm³   [1]

Link the words **A** to **F** on the left to their meanings **1** to **6** on the right.

| A Deoxygenation | 1 Bacteria, viruses and microscopic animals |

| B Heavy metals | 2 Removal of oxygen especially from water or other liquids |

| C Microbes | 3 Elements such as mercury, lead and copper which are poisonous |

| D Phosphates | 4 Waste solid and liquids produced by humans |

| E Sewage | 5 Poisonous |

| F Toxic | 6 Compounds found in most fertilisers which contain phosphorus |

[3]

1. Complete these sentences about plastic pollution in water using words from the list.

fish   **liver**   **microplastics**   **rivers**   **small**   **trap**

Plastics can get into the oceans from ships, coastal towns and by transport in .... from

inland. Plastic fishing nets can .... or kill .... and other sea creatures. Very

small particles of plastics called .... have been found in drinking water. These particles

are so .... that they can get into our bloodstream and then to organs such as the

.... and kidney, where they may cause harm. [6]

2. Link the start of these sentences about water pollutants **A** to **D** to their affects **1** to **4**.

| A Heavy metals .... | 1 .... are generally unreactive but can harm aquatic life. |

| B Nitrates .... | 2 .... contains harmful microbes which cause disease. |

| C Plastics | 3 .... cause the deoxygenation of water by the process of eutrophication. |

| D Sewage | 4 .... such as mercury are toxic. |

[2]

Language lab

Link the names **A** to **D** on the left with the descriptions **1** to **4** on the right.

A Alcohols	**1** A group of organic compounds having a – COOH group
B Alkanes	**2** Hydrocarbons having one or more C = C bonds
C Alkenes	**3** A group of organic compounds having an – OH group
D Carboxylic acids	**4** A group of hydrocarbons with only single bonds

[2]

1. **a.** State the meaning of the term homologous series.

   ....

   .... [2]

   b. To which homologous series do these compounds belong?

   Propene ....   Butanol ....

   Hexane ....   Propanoic acid .... [4]

2. Write the molecular formula for each of these compounds.

   Ethane ....   Ethanol ....

   Ethanoic acid ....   Ethene .... [4]

3. Write the full displayed formula for each of these compounds showing all atoms and bonds.

a. Methanol, CH_3OH	b. Methanoic acid, HCOOH

   [2]

4. Write the general formulae for:

   a. Carboxylic acids ....   **b.** Alkanes ....

   c. Alkenes ....   **d.** Alcohols .... [4]

Language lab

Link the names **A** to **D** on the left with the descriptions **1** to **4** on the right.

A Alkyl group	**1** A compound that contains only carbon and hydrogen atoms
B Displayed formula	**2** This shows the number and type of each atom in one molecule of a compound
C Hydrocarbon	**3** This is shows all of the atoms and all of the bonds
D Molecular formula	**4** This is formed by removal of a hydrogen atom from an alkane

[2]

1. Write the displayed formula for each of these compounds.

a. Propane, $CH_3CH_2CH_3$	**b.** Propanol, $CH_3CH_2CH_2OH$
c. Butene, $CH_3CH_2CH=CH_2$	**d.** Ethanoic acid, CH_3COOH
e. Ethene, $CH_2=CH_2$	**f.** Pentane, $CH_3CH_2CH_2CH_2CH_3$

[6]

2. Link each of the alkyl groups in **a.** to **d.** with one of these formulae. (Note: not all the formulae are used.)

$$CH_3-　　C_2H_5-　　C_3H_7-　　C_4H_9-　　C_5H_{11}-　　C_6H_{13}-$$

a. propyl ....　　**b.** methyl ....

c. hexyl ....　　**d.** ethyl ....　[4]

Complete these sentences about structural formulae and isomers using words from the list.

atoms   carbon   isomers   molecular   number   structural   unambiguous

A structural formula is the simplest .... (clearly defined) description of how the ....

are arranged in a molecule. It shows each .... atom in a molecule together with the ....

of the other atoms attached to each carbon atom. Compounds with the same .... formula but

different .... formulae are called ....   [7]

1. Write the structural formula of each of these compounds on the dotted line beneath each structure.

a.

```
    H   H   H   H
    |   |   |   |
H — C — C — C — C — H
    |   |   |   |
    H   H   H   H
```

....

b.

```
    H   H
    |   |
H — C — C — O — H
    |   |
    H   H
```

....

c.

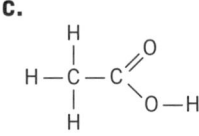

....

d.

```
    H   H
    |   |      H
H — C — C = C
    |          \
    H           H
```

....

e.

```
    H   H   H
    |   |   |
H — C — C — C — H
    |   |   |
    H   O   H
         \
          H
```

....

f.

```
    H   H   H
    |   |   |
H — C — C — C — H
    |   |   |
    H   H   H
```

....   [6]

2. **a.** Draw the displayed formula of the compound with the structural formula $CH_3CH_2CH_2CH_2CH_3$.

.... [1]

   b. Draw the displayed formulae of **two** other isomers of this compound.

.... [2]

3. Draw two isomers of the alkene with the molecular formula, C_4H_8.

.... [2]

Use the clues below to do this crossword

Across: 4. A liquid which vaporises easily is described as ....

5. A product of combustion.

6. Symbol for one of the elements in a compound used to test for water.

7. Fuel used for lorries and cars

9. The physical state of coke at r.t.p.

Down: 1. Black solid fossil fuel

2. The main compound in natural gas

3. Another name for gasoline

4. A liquid which is thick and sticky is described as ....

8. Petroleum is separated into fractions in an oil refin...........

[10]

1. Petrol can be made from coal using the following route:

| A coal | → | B decomposition using heat and hydrogen | → | C refining based on boiling point | → | D petrol |

a. Which stage, **A**, **B**, **C** or **D** involves distillation? .... [1]

b. Which stage involves reduction? .... [1]

2. The table shows some properties of different petroleum fractions.

fraction	boiling point range / °C	size of molecules	volatility	viscosity	ease of burning
1	up to 100				
2	100–150				
3	150–200				
4	200–300				

a. In the third column draw an arrow to show how the size of the molecules varies with the boiling point range (low → high) [1]

b. i. Some compounds are volatile. What is the meaning of the term volatile?

.... [1]

ii. In the fourth column draw an arrow to show how the volatility of the compounds varies with the boiling point range (volatility low → high) [1]

c. In the fifth column draw an arrow to show how the viscosity of the compounds varies with the boiling point range (viscosity low → high) [1]

d. In the sixth column draw an arrow to show how the ease of burning of the compounds varies with the boiling point range (difficult→ easy) [1]

Complete these sentences about the distillation of petroleum using words from the list.

boiling   bottom   condense   further   higher   lower   temperatures   top

There is a range of .... in the distillation column, hot at the .... and cooler at

the .... Hydrocarbons with .... boiling points move .... up the column

and .... when the temperature in the column falls just below the .... point of the

hydrocarbons. Hydrocarbons with .... boiling points condense lower down the column.   [8]

1. The diagram shows a simple experiment to separate petroleum fractions.

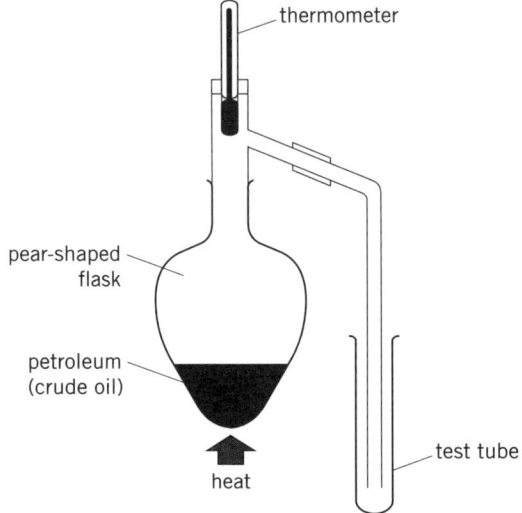

thermometer

pear-shaped flask

petroleum (crude oil)

heat

test tube

Describe how you could use this apparatus to separate the hydrocarbon fractions.

....

....

....

....   [4]

2. Link the petroleum fractions **A** to **D** on the left with their uses **1** to **4** on the right.

A Bitumen	1 Making chemicals
B Fuel oil	2 Jet fuel
C Kerosene	3 Fuel for home heating and ships
D Naphtha	4 Making road surfaces

[2]

Language lab

Complete these sentences about the reaction of alkanes with chlorine.

chlorine	combustion	excess	green	hydrogen

photochemical	substitution	sunlight	unreactive

Alkanes are generally .... except for the reaction with chlorine in the presence of

light (a .... reaction) and .... When .... alkane is mixed

with chlorine in a sealed tube and exposed to ...., the .... colour of the

chlorine disappears. A chlorine atom replaces a .... atom in the alkane. This type of

reaction is called a .... reaction. If excess .... is present, more than one

hydrogen atom is replaced by chlorine. [9]

1. Complete these sentences about alkanes.

   a. Alkanes are .... because they only have hydrogen and carbon atoms in their structure. [1]

   b. All the bonds in alkanes are .... bonds. [2]

2. How do the boiling points of the alkanes change with increase in relative molecular mass?

   .... [1]

3. Complete these equations for the typical reactions of alkanes.

   a. $C_5H_{12} +$ .... $O_2 \rightarrow$ .... $CO_2 +$ .... H_2O [2]

   b. .... $C_4H_{10} +$ .... $\rightarrow$ .... $CO +$ .... H_2O [2]

   c. $CH_4 + Cl_2 \rightarrow$ .... $+ HCl$ [1]

   d. $C_2H_6 +$ .... $Cl_2 \rightarrow C_2H_4Cl_2 +$ .... [2]

4. Which two words describe the reaction in **3.c**. Put rings around the correct answers.

   addition   cracking   neutralisation   photochemical   polymerisation   substitution [2]

5. Draw the two isomers of the hydrocarbon with the formula C_4H_{10} showing all atoms and all bonds.

[2]

Language lab

Use the clues below to do this crossword.

Across: 1. Breakdown of alkanes into smaller alkanes and alkenes by heat

3. Alkene formed by cracking ethane.

5. A gas formed by cracking which is not a hydrocarbon

7. Cracking is a type of .... decomposition

Down: 1. This speeds up cracking

2. $C_{16}H_{34} \rightarrow A + C_7H_{14}$
How many C atoms does A have?

4. A condition needed for cracking

6. Aluminium .... is a catalyst used in cracking

[8]

1. The bar chart shows the supply and demand for different petroleum fractions.

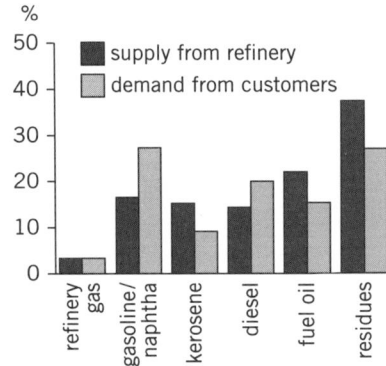

a. Which of the fractions shown has molecules with the longest chains?

.... [1]

b. Name and give the formula of one molecule present in refinery gas.

.... [2]

c. i. For which fractions is the demand much greater than the supply?

.... [1]

ii. For which fractions is the supply much greater than the demand?

.... [1]

2. Complete these equations for cracking.

a. $C_{10}H_{22} \rightarrow C_4H_{10} + \ldots\ldots$ [1]

b. $C_{14}H_{30} \rightarrow C_3H_8 + C_4H_8 + \ldots\ldots$ [1]

Complete these sentences about saturated and unsaturated compounds using words from the list.

aqueous   decolourised   orange   remains   saturated   unsaturated

We can tell the difference between an unsaturated and a .... compound by adding

.... bromine to a sample of the compound. Aqueous bromine is .... in

colour. If the aqueous bromine is ...., the compound is .... If the

aqueous bromine .... orange, the compound is saturated.   [6]

1. **a.** The table gives some information about the alkenes. Complete the table in the spaces provided. Note that the boiling points only need be approximate.

name of alkene	molecular formula	boiling point / °C
ethene	....	−102
....	C_3H_8	....
....	....	−7
pentene	....	30
hexene	C_6H_{12}	....

[7]

    b. Which of the alkenes in the table are likely to be liquids at r.t.p.?

.... [1]

2. The structure of compound **T** is shown below.

$$H_2N-CH=CH-CH_2-O-H$$

On the structure of **T** above draw a ring around the functional group which makes this compound unsaturated.   [1]

3. Complete the boxes in the diagram below to show the structure of **A** and formula of the additional reagent **B** including the state symbol.   [3]

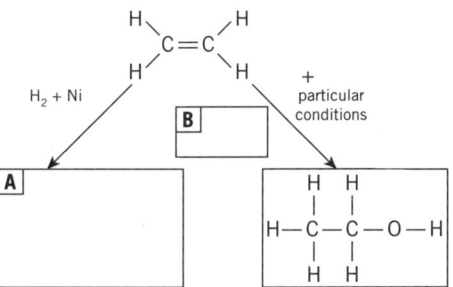

101

Complete these sentences about ethanol using words from the list.

acidified   air   bacteria   dioxide   catalyse   ethanoic   liquid   manganate(VII)   water

Ethanol is a colourless .... at r.t.p. It burns in excess .... to produce

carbon .... and .... Ethanol can be oxidised to .... acid

by oxygen from the air in the presence of .... Enzymes from the bacteria

.... the reaction. In the laboratory, ethanol can be converted to ethanoic acid

by oxidation with .... potassium ....   [9]

1. The diagram shows the apparatus used to convert ethanol to ethanoic acid.

   water out

   water in

   ethanol + potassium manganate(VII) + acid

   a. On the diagram, draw an arrow to show where heat is applied. [1]

   b. What type of reagent is the potassium manganate(VII)?

   .... [1]

   c. State the colour change of the potassium manganate(VII) when not in excess.

   from ....

   to .... [2]

   d. Why is the condenser in the upright position?

   ....

   ....

   .... [2]

2. a. Write a chemical equation for the complete combustion of ethanol.

   .... [2]

   b. Complete the equation for the oxidation of ethanol to ethanoic acid.

   .... + 2[O] → .... + ...   [2]

3. Give two uses of ethanol.

   .... **and** .... [2]

Language lab

Search for the names of six words about the production of ethanol by fermentation, including names of reactants and products. Words go forwards or downwards only. Write the names in the space below.

X	E	T	H	A	N	O	L	A
G	L	U	C	O	S	E	Y	S
C	E	Q	Y	W	E	R	T	D
A	N	A	E	R	O	B	I	C
R	Z	Z	A	V	X	C	V	F
U	Y	W	S	A	T	N	B	A
O	M	E	T	F	Y	U	I	Y
N	E	A	S	D	F	G	H	J
L	S	R	D	I	S	T	I	L

[6]

1. **a.** Ethanol can be manufactured by fermentation or by hydration of ethene.

   Complete the table about these reactions.

	fermentation	hydration of ethene
Reagents needed		
Temperature / °C		
Pressure		
Catalyst		

   [8]

   b. Give two disadvantages of producing ethanol by fermentation.

   ....

   .... [2]

   c. Give two advantages of producing ethanol by fermentation.

   ....

   .... [2]

   d. Give one disadvantage of producing ethanol by hydration of ethene.

   .... [1]

   e. Give two advantages of producing ethanol by hydration of ethene.

   ....

   .... [2]

103

Language lab

Link the names of the esters **A** to **D** on the left to their formulae **1** to **4** on the right.

| **A** Butyl ethanoate | **1** $HCOOCH_3$ |

| **B** Ethyl propanoate | **2** $CH_3COOCH_2CH_2CH_3$ |

| **C** Methyl methanoate | **3** $CH_3COOCH_2CH_2CH_2CH_3$ |

| **D** Propyl ethanoate | **4** $CH_3CH_2COOCH_2CH_3$ |

[2]

1. Ethanoic acid dissolves in water.

$$CH_3COOH + H_2O \rightleftharpoons CH_3COO^- + H_3O^+$$

a. How does this equation show that ethanoic acid is a weak acid?

....

.... [2]

b. Explain why water is acting as a base in the forward reaction.

.... [2]

2. Complete these equations for some reactions of ethanoic acid.

a. .... CH_3COOH + ... Na → .... + ...   [3]

b. .... CH_3COOH + Mg → .... + ...   [3]

c. CH_3COOH + NaOH → .... + ...   [2]

d. .... + ... → CH_3COOCH_3 + ...   [3]

3. Draw the structure of these esters showing all atoms and all bonds.   [2]

a. ethyl butanoate	**b.** propyl methanoate

4. Name these esters:

a. $HCOOC_4H_9$ .... **b.** $C_2H_5COOC_3H_7$ .... [2]

Language lab

Complete these sentences about polymers using words from the list.

bonds   ethene   join   molecule   monomers   polymerisation

A polymer is a substance which has very large .... formed when lots of

small molecules called .... join together. This process is called .... When

poly(ethene) is formed, one of the C=C .... of .... is broken and

the monomers .... together in a chain.   [6]

1. In the box below draw a section of the polymer chain formed by the addition of four units of ethene monomers.

[2]

2. Many plastics are non-biodegradable.

    What is meant by the term *non-biodegradable*?

    ....

    .... [2]

3. We can get rid of waste plastics by recycling, putting them in landfill sites or by burning them.

    a. Give one advantage of recycling plastics.

        .... [1]

    b. Give one disadvantage of burning plastics.

        .... [1]

4. Complete these sentences about recycling the polymer, PET, using words from the list.

    aqueous   hydrolysed   monomers   repolymerised

    PET polymers are .... (broken down using .... acids or alkalis) to their ....

    The monomers are then ....   [4]

105

Language lab

Complete these sentences about addition polymers using words from the list.

addition   bonds   combine   formed   monomers   no   two

When .... containing C=C double .... are polymerised, no other molecule apart

from the polymer is .... We call this type of polymerisation .... polymerisation.

An addition reaction is a reaction where .... or more molecules ....

and .... other molecule is formed.   [7]

1. The structure of a monomer is shown below.

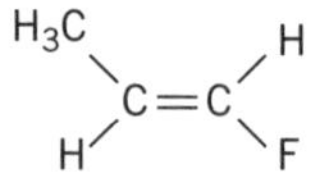

$$H_3C \diagdown \diagup H$$
$$C=C$$
$$H \diagup \diagdown F$$

Draw a section of the polymer chain formed from this monomer. Show three repeat units.

[3]

2. Draw the structure of the polymer formed from but-2-ene, $CH_3–CH=CH–CH_3$, as one repeat unit of this polymer with brackets and *n*.

[3]

3. Draw the monomers of polymers **A** and **B**.

Polymer **A**   Polymer **B**

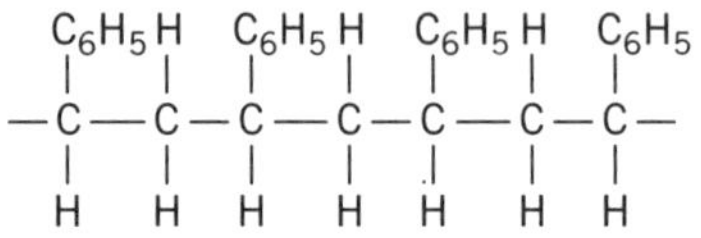

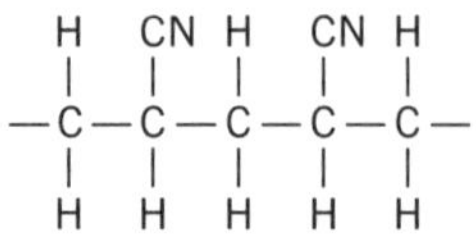

$$\begin{array}{ccccccc} C_6H_5 & H & C_6H_5 & H & C_6H_5 & H & C_6H_5 \\ | & | & | & | & | & | & | \\ -C & -C & -C & -C & -C & -C & -C- \\ | & | & | & | & | & | & | \\ H & H & H & H & H & H & H \end{array}$$

$$\begin{array}{ccccc} H & CN & H & CN & H \\ | & | & | & | & | \\ -C & -C & -C & -C & -C- \\ | & | & | & | & | \\ H & H & H & H & H \end{array}$$

[4]

Complete these sentences about condensation polymerisation using word from the list.

chloride   eliminated   functional   small   water

In condensation polymerisation, molecules with different .... groups react together. In addition

to the polymer a .... molecule such as .... or hydrogen ....

is .... [5]

1. The diagram shows two polymers, **A** and **B**.

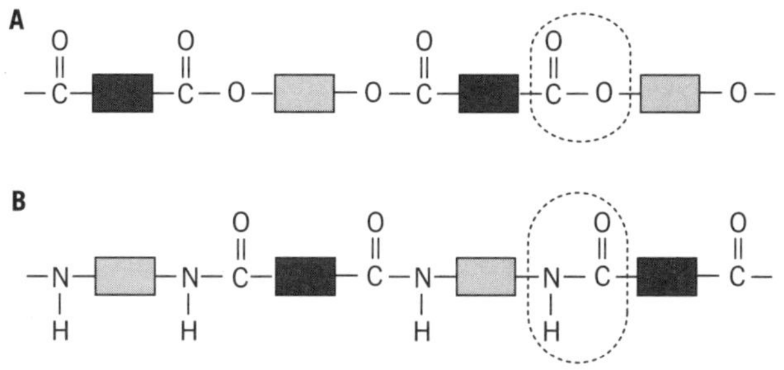

A

B

a. On each of the diagrams above put brackets to show one repeat unit. [2]

b. Give the names of the linkages shown within the dotted circles in the diagram.

   i. Polymer **A** .... ii. Polymer **B** .... [2]

2. The structure of poly(lactic acid) is shown below.

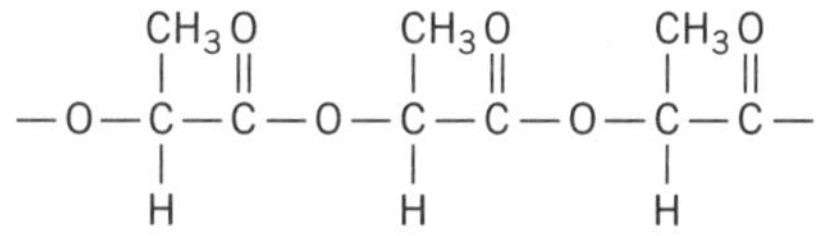

a. Give the names of the two functional groups which react to form this polymer.

   .... and .... [2]

b. Draw the structure of the single monomer which is used to form this polymer.

[2]

107

Language lab

Complete these sentences about proteins using words from the list.

amide **amino** **condensation** **different** **order** **repeat** **twenty**

Proteins are .... polymers. They have the same .... linkage as nylon-6,6 but instead of being formed from two .... monomers, proteins are made from about .... different .... acids. There is no regular .... unit in proteins because the amino acid residues are not in a regular .... [7]

1. The diagram below shows part of a protein.

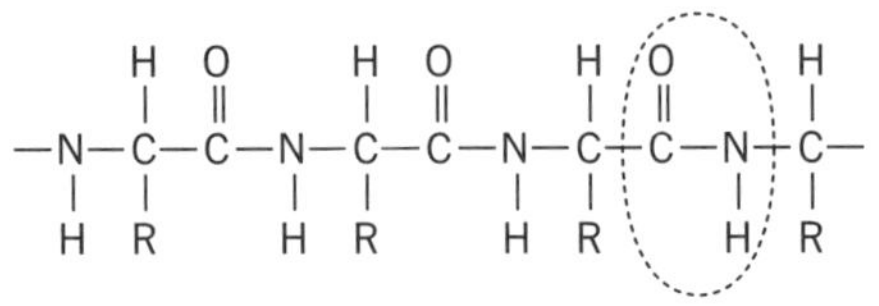
H O H O H O H
N C C N C C N C C N C
H R H R H R H R

a. On the diagram above, put brackets to show one amino acid residue. [1]

b. Give the name of the linkage CONH. .... [1]

2. The structure of part of poly(glycine) is shown below

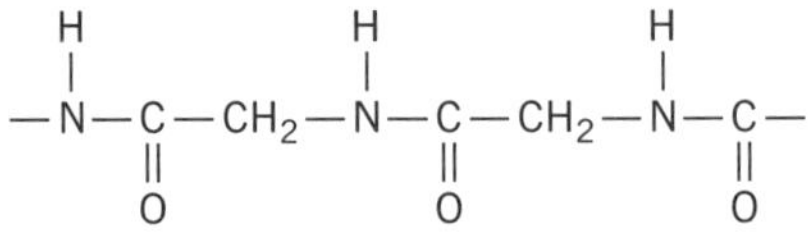
H H H
N C CH2 N C CH2 N C
O O O

Deduce the structure of the monomer which is used to form this polymer.

[2]

3. Proteins are hydrolysed to amino acids. What is the meaning of the term *hydrolysis*?

.... [2]

Some definitions in chemistry involve two or more different points. This exercise will help you remember how to write these definitions.

1. **a.** Complete this definition of the term *hydrocarbons* using words from the list.

    carbon   **compounds**   **only**

    Hydrocarbons are .... containing .... hydrogen and

    .... atoms.   [2]

    b. Which word do you think is often left out by students when defining a hydrocarbon?

    .... [1]

2. **a.** Complete this definition of the term *compound* using words from the list.

    bonded   **different**   **substance**   **two**

    A compound is a .... which contains .... or more .... atoms

    .... together.   [2]

    b. Which two words do you think are often left out by students when defining a compound?

    .... [2]

3. **a.** Complete this definition of the term *relative atomic mass* using words from the list.

    average   **element**   **isotopes**   **mass**   **twelfth**

    Relative atomic mass is the .... mass of the .... of an ....

    compared to one- .... of the .... of an atom of ^{12}C.   [3]

    b. Check the answer to see if you are correct, then write out the definition again and underline the five most important words in the definition. They are not all the same as the gaps!

    ....

    ....

    .... [3]

4. Complete this definition of the term mole using words from the list.

    amount   **constant**   **molecules**   **particles**

    A mole is the .... of substance that contains 6.02×10^{23} specified ....

    (atoms, ions, ... or electrons). This number of particles is called the Avogadro ....   [2]

109

Some words are essential when describing chemical terms and writing extended answers (answers requiring writing in phrases or sentences). This exercise will help you remember these words.

In each of these sentences, cross out the incorrect word option and / or write in the missing words.

1. Activation energy is the <u>maximum / minimum</u> energy that colliding particles must have for a reaction

    to take place.   [1]

2. An element is a substance made up of only one .... of atom.   [1]

3. Isotopes are <u>atoms / molecules</u> of an <u>compound / element</u> with the same number of protons but different

    numbers of neutrons.   [2]

4. Electrolysis is the .... of an ionic compound, when molten or in aqueous solution, by

    the passage of ....   [2]

5. An oxidising agent oxidises another substance by <u>gaining / losing</u> electrons.   [1]

6. The conduction of electricity in an electrolyte is due to the movement of <u>electrons / ions</u>.   [1]

7. When hydrochloric acid reacts with magnesium, increasing the concentration of the acid ....

    the rate of the reaction because there are more particles of acid in a given <u>area / volume</u>. So the

    <u>frequency / number</u> of collisions increases.   [3]

8. Sodium chloride conducts electricity when molten because the <u>atoms / ions</u> are free to .... [2]

9. When sodium reacts with chlorine, each sodium <u>atom / ion</u> loses an electron. Each chlorine <u>atom / molecule</u>

    gains one electron.   [2]

10. The <u>atoms / components</u> of a mixture can be separated by <u>chemical / physical</u> means.   [2]

11. Unsaturated compounds contain carbon–carbon <u>double / single</u> bonds. When added to aqueous bromine

    the colour of the aqueous bromine changes from <u>orange / red</u> to <u>clear / colourless</u>.   [3]

1. Link each of the words **A** to **F** about the properties of some metals to the meanings **1** to **6**.

A Dense	**1** Can be beaten into shape
B Ductile	**2** Not easily changed by a force
C Lustrous	**3** Can be drawn into wires
D Malleable	**4** Rings when hit
E Sonorous	**5** Has high mass to volume ratio
F Strong	**6** Shiny

[3]

2. Link each of the terms **A** to **G** about environmental chemistry to the phrases **1** to **7**.

A Acid rain	**1** Can be broken down by organisms
B Adverse effect	**2** A gas which warms the atmosphere
C Biodegradable	**3** Burning in limited supply of air
D Climate change	**4** Thick mist caused by NO_2 and hydrocarbons
E Greenhouse gas	**5** An example of this is desertification
F Incomplete combustion	**6** Harmful action
G Photochemical smog	**7** Precipitation which has pH 5 or less

[4]

3. Link each of the chemical terms **A** to **F** to the meanings **1** to **6**.

A Decomposition	**1** Reaction of acid with a base to form a salt and water
B Displacement	**2** Loss of electrons
C Distillation	**3** Reaction which depends on light
D Neutralisation	**4** One atom or group replacing another
E Photochemical	**5** Separation due to different boiling points
F Reduction	**6** Breakdown of a substance

[3]

1. Link the beginnings of these sentences **A** to **G** with the correct endings **1** to **7**.

A We can use the titration method to	**1** ..... separate a mixture of liquids with different boiling points.
B We use upward displacement of air to	**2** ..... make an insoluble salt from two soluble compounds.
C Fractional distillation is used to	**3** ..... collect gases which are insoluble in water.
D We can use downward displacement of water to	**4** ..... deliver a given volume of liquid accurately.
E We can use the precipitation method to	**5** ..... make a soluble salt from an alkali and an acid.
F A burette is used to	**6** ..... separate a solid from a solution.
G Filtration is used to	**7** ..... collect gases which are denser than air.

[4]

2. Join up these fragments to form a sentence describing why diamond has a high melting point.

    ..... between all the carbon atoms.   Diamond has a high melting point .....

    ..... covalent bonds which exist .... to break the strong .....

    ..... because it takes a lot of energy .....

    ....

    .... [2]

3. Join up these fragments to form two sentences describing how the rate of reaction changes with temperature.

    ..... the faster the particles move .... reactant particles being successful.

    ..... equal to or greater than the activation energy .... also increases, so there is .....

    The higher the temperature, ...   The number of particles having energy .....

    ..... because they have more energy and collide .... with a greater frequency.

    ..... more chance of collisions between .....

    The higher the temperature, ...

    ....

    ....

    .... [3]

At the beginning of a question is a part which is called the 'stem'. This tells you what the question is about.

- The stem may include diagrams or tables.
- Look for key words in the stem which tell you about the question.
- You might find it useful to underline key words.
- Make sure that you read every word. Sometimes it is the non-scientific words which cause trouble because when we are reading quickly, we sometimes miss these out.
- Remember that you may need to refer back to the stem of the question when you answer different parts of the question.

The diagram below shows you a question stem and what to look for.

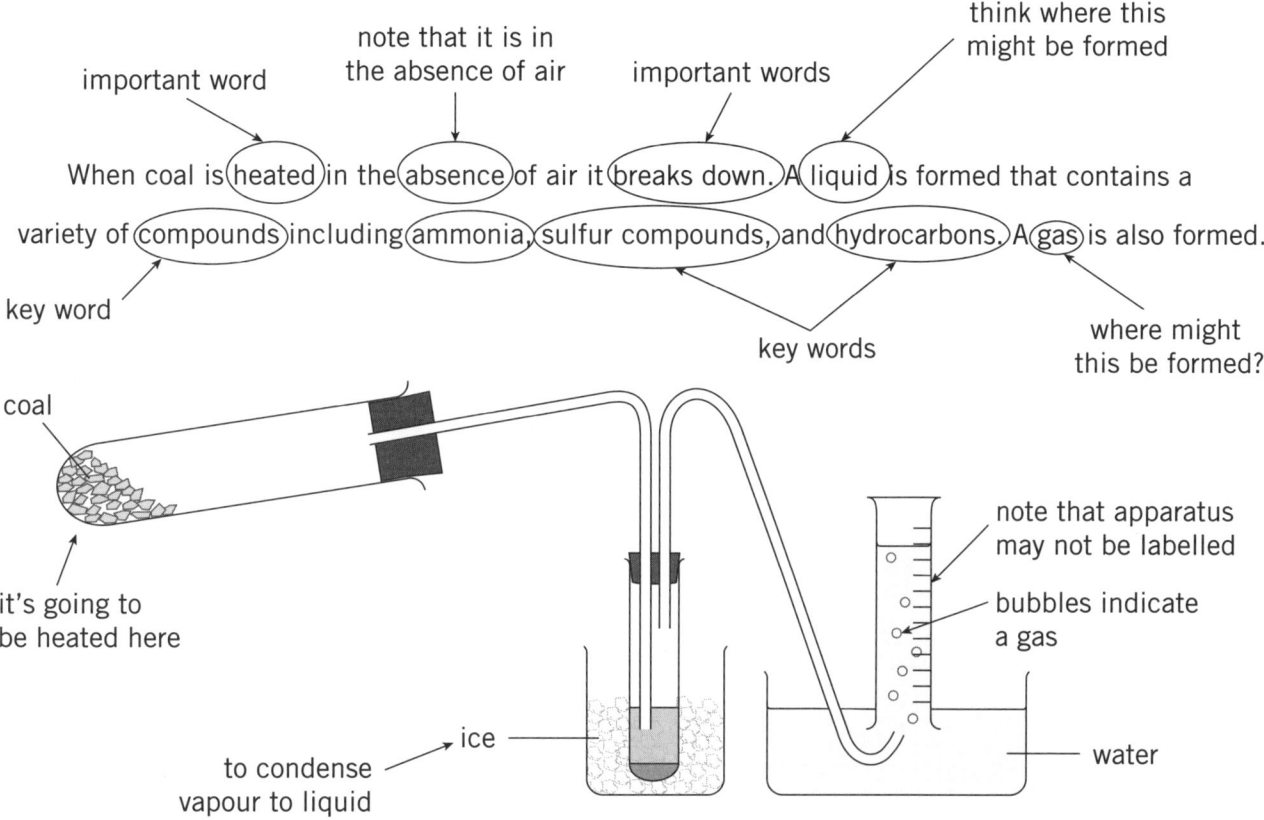

What can you find out from the stem and diagram without any questions being asked?

- It's a thermal decomposition reaction because of the words heat and breaks down (decomposition means breaks down).
- A gas is collected in the measuring cylinder (you need to know the names of pieces of apparatus).
- The gas is insoluble in water (otherwise it would dissolve in the water and not collect at the top).
- Its the coal that is heated. So if you are asked to draw an arrow to show where the heat is applied it should be under the coal not anywhere else.

When reading each part of a question:

- Read each word carefully, including the non-scientific words. Don't rush through the passage.
- Look for key words including command words. A list of the most important command words is given in Units 22.1 and 22.2.
- Note the number of examples that you have to give.

1. Underline the most important words in each of these questions. The number of words to underline is given after each example.

    a. Complete the reaction pathway diagram to show reactants, products and the enthalpy change. (Underline 4 words)   [4]

    b. Describe how to separate the different coloured compounds present in a mixture of dyes. (Underline 3 words)   [3]

    c. Describe the observations when dilute hydrochloric acid is added to zinc. (Underline 4 words)   [4]

    d. Describe and explain condensation in terms of the energy and movement of particles. (Underline 5 words)   [5]

2. Read these questions and the incorrect answers given.
    In each case describe and explain what mistake has been made in reading the question.

    a. Question: Give two properties shown by all metals.

        Incorrect answer: high density, high melting point

        Mistake: ....

        .... [2]

    b. Question: Give one observation when nitric acid is added to zinc?

        Incorrect answer: hydrogen gas is given off

        Mistake: ....

        .... [2]

    c. Question: Define a strong acid.

        Incorrect answer: hydrochloric acid

        Mistakes: ....

        .... [3]

    d. Question: What feature of the ethene molecule shows that it is an unsaturated hydrocarbon?

        Incorrect answer: It decolourises bromine water

        Mistake: ....

        .... [2]

Command words tell us what sort of thing we need to write in response to a question. Here is a list of the command words used in chemistry and how to respond to them.

Calculate: You need to work out a problem using numbers.
- The problem may be in several stages.
- Look out for the number of marks: This often shows you the number of stages needed in the calculation.
- Always show your working.

  Example: Calculate the mass of 11 dm^3 of carbon dioxide. The answer involves **i.** finding the moles of carbon dioxide using the relationship that 1 mol of a gas occupies 24 dm^3 then **ii.** multiplying moles by the molar mass of carbon dioxide.

Compare: You have to comment on similarities and differences.

  Example: Compare the properties of the transition elements and the Group I elements.

Deduce: Conclude from available information.

  Example: Sodium sulfate is Na_2SO_4. Deduce the formula of the sulfate ion. Answer: SO_4^{2-}

Define: Give a precise meaning. You need to write down the main points about a term.
- The only way to do this is by memorising key terms.

  Example: Define oxidation in terms of electron transfer. Answer: oxidation is loss of electrons.

Describe: State the main points about something. You may have to write about a sequence of events, draw a diagram or state what happens.
- If you are asked to describe observations, remember to state what you see, hear, smell or feel. Do not describe the names of the substances.

  Example 1: Describe how to obtain sodium chloride crystals from a solution of sodium chloride. Answer: Warm to the crystallisation point then filter off the crystals.

  Example 2: Describe what happens when you add acid to an aqueous solution of sodium carbonate. Answer: Bubbles are seen. NOTE: The answer 'carbon dioxide is given off' is not correct.

Determine: The answer can be obtained from a graph, by calculation or other information.

  Example: Determine the value of the gas released after 20 seconds.

Evaluate: Write something about the quality, importance or value of something. It often involves being critical.

  Example: Evaluate the experiment by referring to the apparatus and the range of results.

Explain: You have to use particular ideas to describe why something happens.

  Example: Explain why the volume of a gas increases with temperature. Answer: The particles of gas move faster and get further away from each other (use of kinetic particle theory).

Identify: Name, select or recognise something from given information.

   Example: Identify the reducing agent in this reaction.

Justify: Support a case with evidence or ideas.

   Example: Justify the position of silicon in the Periodic Table using its electronic configuration and the information about the physical properties of other elements in Group IV given.

Predict: You have to make connections between various items of data.

• You often have to extrapolate or interpolate data when answering these questions.

   Example: Predict the melting point of potassium (when given the melting points of other Group I elements). The answer involves looking at the melting points of the elements either side of potassium and choosing a suitable value in between these.

Show (that): Give structured evidence that leads to a result.

   Example: Show by calculation that calcium carbonate is the limiting reactant.

State / Give: Produce a short answer from a given source or memory.

   Example: State / Give the electronic structure of sodium. Answer: 2,8,1.

Suggest: You have to use your general chemical knowledge to write about a situation which is unfamiliar to you. You may need to:

• Think of similar substances to the one which is being asked about.
• Think of general ideas of structure, bonding, electrolysis, redox, rate or equilibrium.

   Example: The structure of boron nitride is similar to graphite. Suggest why boron nitride is slippery. Answer: Think about the properties of graphite that make it slippery, then repeat these for boron nitride, e.g. weak forces between the layers, layers can slide over each other.

NOTE: Command words are often combined:
   Example: State the meaning of the term combustion.
   Describe and explain the effect of increasing the temperature on the position of this equilibrium.

1. Underline the command words in each of these questions.
   a. Use the information in the table above to suggest a value for the boiling point of propane.   [1]
   b. Describe how distillation is carried out and give the name of the physical property on which it is based.   [2]
   c. Use this information to deduce the formula of this oxide of tin.   [1]
   d. Describe the process of diffusion and explain this process in terms of the kinetic particle theory.   [2]
   e. Draw a graph of volume of carbon dioxide against time to determine the volume of carbon dioxide formed in the first 30 seconds.   [2]

You should find the best way of revising for yourself. There is no one correct way of revising.

- Don't just read through books or notes and hope that you will remember things.
- Don't leave revision to the last moment. It is best to revise material throughout the year.
- Revision should be active (see Unit 22.4).

Here is a check list of things to help you find the best way for you to revise:

- Find the best time of day to revise. Some people revise better in the evening, others in the morning.

  When do I revise best? ....

- Find a time when you will not be disturbed.

  When is this most likely to be? ....

- Find the best conditions needed for you to revise. Some people prefer to revise in absolute silence; others find it useful to have some music in the background.

  Is music in the background really good for you when revising? ....

  ....

- Revise regularly. You may find it useful to revise a topic about a week after you have finished it, to make sure that you have really understood it.

  Do you revise only for exams or tests? ....

- Find the best length of time for each revision session. You may find that several short periods of revision, for example 3 spells of 20 minutes with breaks in between are more productive than one longer period of revision.

  What's your best revision span? ....

- Don't imagine that you are revising usefully unless you test yourself from time to time to prove that you are remembering material.

  Do you test yourself? ....

- Make sure that you pay more attention to topics which you find difficult. Don't ignore them!

  What topics in Chemistry do you find difficult? ....

  ....

- Do you remember information better in written form or in the form of diagrams?

  ....

Active revision involves you carrying out different sorts of activities and testing yourself to see if your revision has been successful.

- Revise with others, especially with other classmates. Asking each other questions is a useful way of helping you remember things.
- Test yourself or get someone else to test you.
- Make a list of key areas that you find difficult and concentrate on these.
- Make up mnemonics like OIL RIG for Oxidation Is Loss of electrons and Reduction Is Gain of electrons.
- Look through the syllabus and text book and list general areas which need revision e.g. definitions and key terms to learn e.g. element, isotopes, relative atomic mass. You can test yourself by writing the terms and definitions on a sheet like this:

term	definition
isotopes	Atoms with the same proton number having different numbers of neutrons
element	A substance which contains only one type of atom.
empirical formula	A formula which shows the simplest whole number ratio of the different atoms or ions in a compound.

When you think you have learnt the definitions, cover up or fold over the definition side and see if you can remember it.

1. Give three other general topics throughout the syllabus which could be revised in this way.

   .... [3]

2. On a separate piece of paper write the two columns as shown below.

properties of acids	result
litmus	turns red litmus blue
metals	gives salt + hydrogen

   Complete the table to give at least 4 other properties of acids. [4]

3. Make a table similar to the one above to help you revise the properties of hydrocarbons. [16]

Using past papers

- It is useful to look at the wording in past papers to get a feeling for the language used.
- Look not only for the command words and the smaller words which instruct you what to do.
- Look at the number of marks given for the question. This often gives you an indication of how many different points you need to include in your answer. For simple questions, however, you may need to write two points to get one mark.
- Underline the key words (see Units 21.8 and 22.2).
- Work through calculations and extended exam questions and use the mark schemes if available to see where the marks are awarded.

Mind maps are simplified diagrams which show the main points of a topic in the form of a diagram. They are useful summaries because all the words which get in the way of learning the essential points are removed. They can be made as simple or as complicated as required but it is best to make them simple at the start. Mind maps can also be constructed to show links between different topic areas. An example is shown below.

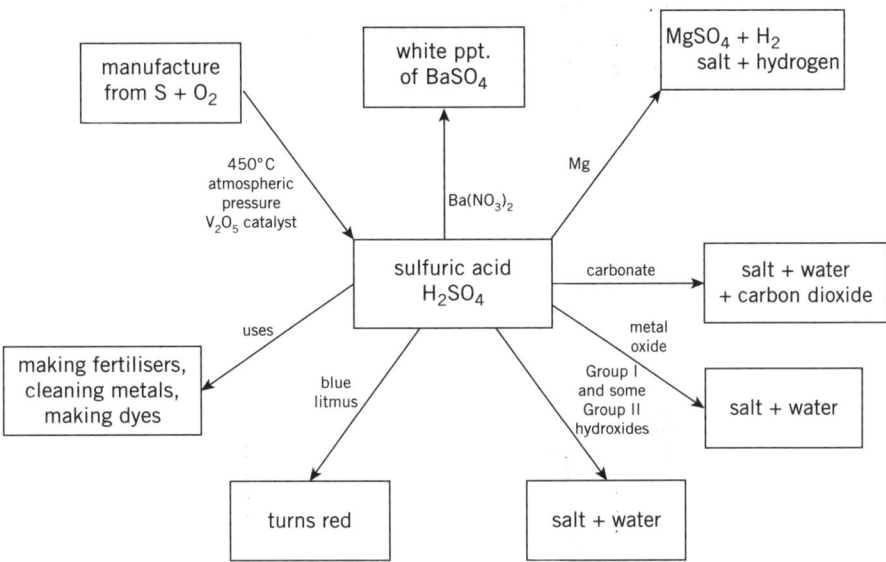

Suggested topics for mind maps which you could make:

Electrolysis; Acids and bases; Methods of purification; Hydrocarbons; Polymers; Properties of the halogens; Iron and steel (or properties of metals); Equilibrium.

In the space below draw a simple mind map (no more than 7 boxes) for a topic of your choice.

1. Complete the mind map for structure by filling in the gaps **A** to **L**.

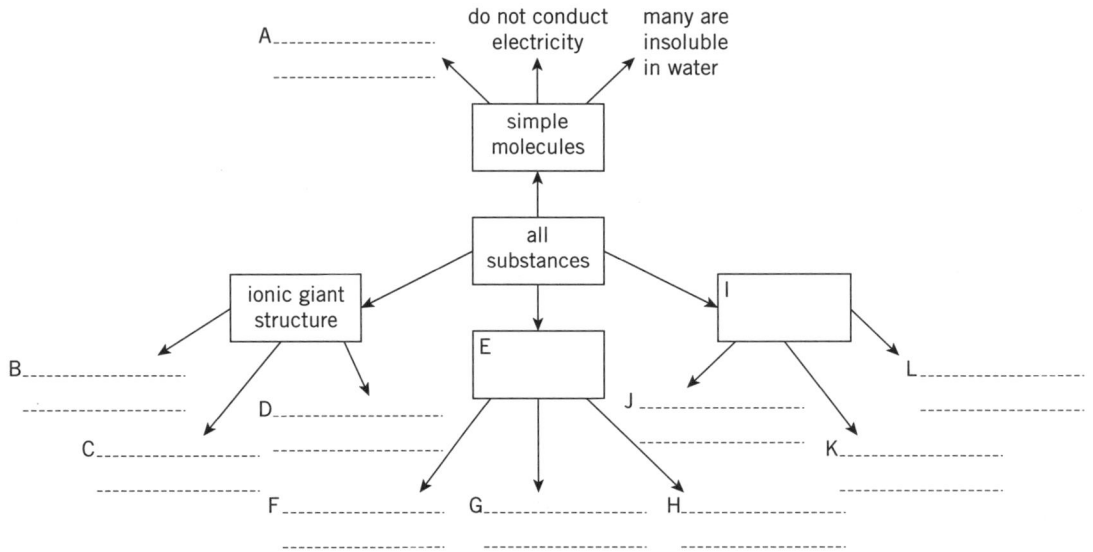

[12]

2. In the space below complete a mind map about rates of reaction.

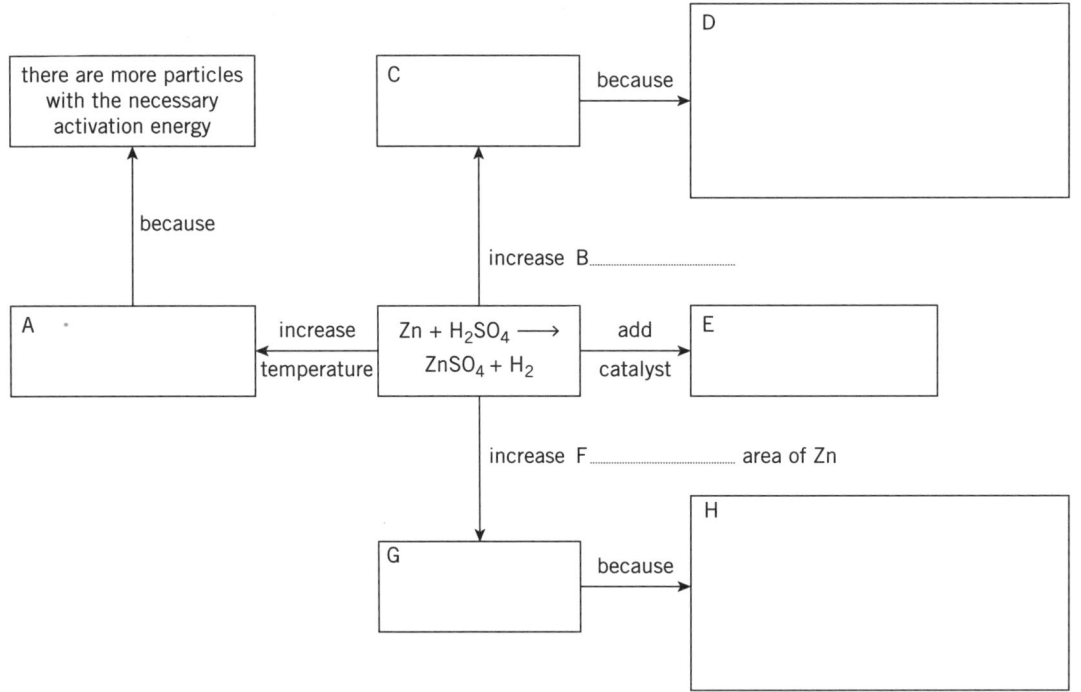

[8]

1. • A small subscript number after an atom or ion refers only to that atom or ion.

So in one formula unit of $CaCl_2$ there is 1 calcium ion and 2 chloride ions and in one formula unit of Al_2O_3 there are 2 aluminium ions and 3 oxide ions.

How many atoms (or ions) of each type are there in:

a. Na_2O .... b. Mg_3N_2 ....

c. PCl_3 .... d. Al_2O_3 ....

e. H_2SO_4 .... [5]

2. • When balancing equations, a large number in front of a formula unit multiplies all the way through.
   • The number can refer to atoms (or ions) or moles of atoms (or moles of ions).

So in $3SO_3$ there are 3×1 sulfur atoms and $3 \times 3 = 9$ oxygen atoms.

How many atoms (or ions) of each type are there in:

a. $2H_2S$ .... b. $5Mg_3N_2$ ....

c. $3Al_2S_3$ .... d. $2Fe_3O_4$ ....

e. $3Li_2CO_3$ .... [5]

3. When calculating relative molecular masses (or relative formula masses)
   • Multiply the number of each type of atom by its relative atomic mass and then
   • Add the products together

Example: Calculate the relative molecular mass of N_2O_5. (A_r values: N = 14, O = 16)

$2N = 2 \times 14 =$ 28
$5O = 5 \times 16 =$ $\underline{80}$
Total $= 108$ Relative molecular mass of $N_2O_5 = 108$

Calculate the relative molecular mass (or relative formula mass) of these compounds.

a. Al_2O_3 (A_r values: Al = 27, O = 16)

.... [1]

b. Na_2CO_3 (A_r values: C = 12, Na = 23, O = 16)

.... [1]

c. $PbSO_4$ (A_r values: Pb = 207, O = 16, S = 32)

.... [1]

1. • Brackets keep particular groups of atoms together. e.g. (NO_3) for nitrates, (OH) for hydroxides.
   • You must not change the numbers within the brackets.
   • A subscript after a bracket multiplies all through the atoms inside the brackets.

   So in $Mg(NO_3)_2$ there is 1 Mg ion and 2 NO_3 ions
   In $2NO_3$ ions there are 2 N atoms and $2 \times 3 = 6$ O atoms.

   And in $4Mg(NO_3)_2$ there are 4 Mg ions, 4×2 N atoms and 4×6 O atoms.

   How many atoms (or ions) of each element are there in:

   a. $Sn(SO_4)_2$ ....

   b. $(NH_4)_2SO_4$ ....

   c. $Ni(ClO_4)_2$ ....

   d. $Ba(IO_3)_2$ .... [4]

2. You need to work out the number of atoms correctly in order to calculate the relative formula masses of these compounds.

   Calculate the relative formula mass of the compounds in question 1.

   a. $Sn(SO_4)_2$

   Relative formula mass =   [1]

   b. $(NH_4)_2SO_4$

   Relative formula mass =   [1]

   c. $Ni(ClO_4)_2$

   Relative formula mass =   [1]

   d. $Ba(IO_3)_2$

   Relative formula mass =   [1]

3. • Water of crystallisation appears as a dot after the main formula.
   • You add this on separately when calculating formula masses.
   e.g. $CuSO_4 \cdot 5H_2O$ contains $[(1 \times 64) + (1 \times 32) + (4 \times 16)] + 5 \times (2 + 16) = 250$

   Calculate the relative formula mass of $CoCl_2 \cdot 6H_2O$
   (A_r values: Cl = 35.5, Co = 59, H = 1., O = 16)

   .... [1]

You should be able to rearrange expressions from first principles rather than having to rely on a 'triangle' to help you.

- The idea is that whatever you do to one side of the equation, you do to the other.

  Example: $\text{moles} = \dfrac{\text{mass}}{M_r}$

- To make mass the subject: multiply both sides by M_r (to cancel M_r on the right)

  $\text{moles} \times M_r = \dfrac{\text{mass}}{\cancel{M_r}} \times \cancel{M_r}$ So $\text{moles} \times M_r = \text{mass}$

- To make M_r the subject: multiply both sides by $mass$ (to cancel mass on the right)

  $\dfrac{\text{moles}}{\text{mass}} = \dfrac{\cancel{\text{mass}}}{M_r \times \cancel{\text{mass}}}$ then turn both sides upside down: $M_r = \dfrac{\text{mass}}{\text{moles}}$

  Now try rearranging these expressions:

1. $\% \text{ yield} = \dfrac{\text{actual yield}}{\text{theoretical yield}} \times 100$

   a. Make actual yield the subject:

   [1]

   b. Make theoretical yield the subject:

   [1]

2. $\text{concentration (in mol/dm}^3) = \dfrac{\text{moles}}{\text{volume (in dm}^3)}$.

   a. Make moles the subject:

   [1]

   b. Make volume (in dm³) the subject:

   [1]

3. Write the formula for density making mass the subject.

   [1]

123

1. Sometimes, when you do calculations, the number will come out as e.g 2.5⁻04 on your calculator. The ⁻04 is called the index or power to the base 10. Numbers like this on your calculator are examples of standard form. We write this 2.5×10^{-4}.

   $1 \times 10^1 = 10$

   $1 \times 10^2 = 10 \times 10 = 100$

   $1 \times 10^3 = 10 \times 10 \times 10 = 1000$

   $2.5 \times 10^3 = 2.5 \times 10 \times 10 \times 10 = 2500$

   a. Write 1 000 000 in standard form .... [1]

   b. Write 7×10^4 in non-standard form .... [1]

   c. Write 3 300 in standard form .... [1]

   d. Write 3.2×10^3 in non-standard form .... [1]

2. Very small numbers can also be written in standard form.

   $1 \times 10^{-1} = 0.1$ or 1/10

   $1 \times 10^{-2} = 0.01$ or 1/100

   $1 \times 10^{-3} = 0.001$ or 1/1000

   $2.5 \times 10^{-3} = 2.5 \times 0.001 = 0.0025$

   a. Write 0.00001 in standard form .... [1]

   b. Write 5×10^{-3} in non-standard form .... [1]

   c. Write 0.0035 in standard form .... [1]

   d. Write 2.5×10^{-1} in non-standard form .... [1]

3. When you multiply numbers in standard form, you simply add the indices (superscripts).

   Example 1: $(2.4 \times 10^3) \times (2 \times 10^2) = 2.4 \times 2 \times 10^{3+2} = 4.8 \times 10^5$

   Example 2: $(5.0 \times 10^{-2}) \times (1.5 \times 10^4) = 5.0 \times 1.5 \times 10^{-2+4} = 7.5 \times 10^2$

   a. What is the product of $(4.0 \times 10^{-3}) \times (3.5 \times 10^2)$ .... [1]

   b. What is the product of $(2.4 \times 10^{-2}) \times (1.5 \times 10^{-3})$ .... [1]

4. When you divide numbers in standard form, you simply subtract the indices.

   Example: $\dfrac{2.4 \times 10^5}{1.2 \times 10^2} = \dfrac{2.4}{1.2} \times 10^{5-2} = 2.0 \times 10^3$

   a. What is the result of $(4.0 \times 10^{-3}) \div (3.5 \times 10^2)$ .... [1]

   b. What is the result of $(2.0 \times 10^2) \div (7.5 \times 10^{-1})$ .... [1]

1. Percentages

In chemistry the result of a smaller number divided by a larger number is multiplied by 100 to get the percentage. You know you've gone wrong in chemical calculations if your percentage yield is greater than 100%!

Example: What is the percentage yield if the actual yield of a product in a reaction is 4.5 g and the theoretical yield is 5.5 g?

$$\% \text{ yield} = \frac{\text{actual yield}}{\text{theoretical yield}} \times 100 = \frac{4.5}{5.5} \times 100 = 82\%$$

Now try these:

a. What is the percentage yield if the actual yield of iron is 20.0 tonnes and the amount expected from the balanced equation is 22.8 tonnes. (Don't forget to multiply by 100!)

% yield = _____ [1]

b. What is the percentage purity of compound **A** if the mass of impure substance in 3.45 kg of **A** is 0.12 kg. (Note that it's the percentage purity not the percentage impurity that is needed!)

% purity = _____ [2]

2. Volumes and areas

The diagram shows how to calculate the area of one side of a cube and the volume of a cube.

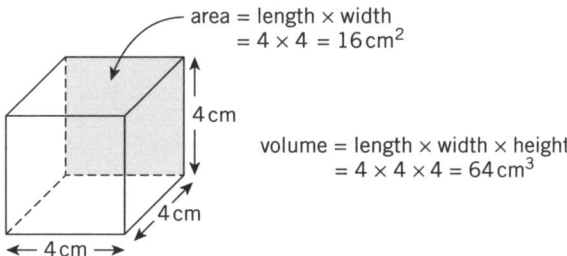

area = length × width
= 4 × 4 = 16 cm²

4 cm

volume = length × width × height
= 4 × 4 × 4 = 64 cm³

4 cm

← 4 cm →

Look at the diagram of the cube above.

a. i. How many sides does the cube have? .... [1]

   ii. What is the total surface area of the cube? .... [1]

b. Calculate the volume of a cube which has a side of 5 cm³.

.... [1]

1. When doing chemical calculations, it is important that we give the answer to the correct number of significant figures and round up figures correctly.

Significant figures:

236.38 has 5 significant figures

32.4 has 3 significant figures

0.0067 has two significant figures (zeros before a number are not significant figures)

0.0300 has 3 significant figures (zeros after a number after a decimal point are significant figures)

Rounding up:

2.3<u>66</u> rounded up to two significant figures is 2.4

2.5<u>57</u> rounded up to two significant figures is 2.6

2.3<u>46</u> rounded up to two significant figures is 2.3

You can see that when rounding if the next figure along is 5 or above, then the figure to be rounded goes up by 1.

a. Round these values to 3 significant figures:

   i. 4.357 ....

   ii. 0.08732 ....

   iii. 137.2 ....

   iv. 0.005498 .... [4]

b. Round these values to 2 significant figures:

   i. 436 ....

   ii. 3.447 ....

   iii. 56.79 ....

   iv. 0.00545 .... [4]

2. When performing a calculation in several stages **do not round up between the steps**. You should only round up at the end. You should round up to the same number of significant figures as the data in the question.

To see the effect of rounding in the middle of a calculation, work through this example.

When 1 mole of pentane is burnt in excess air 5 moles of carbon dioxide are formed. Calculate the volume of carbon dioxide when 6.67 g of pentane burn.

calculation	answer when keeping the figures in your calculator	value when rounding
moles pentane $= \dfrac{6.67}{72.0}$	....	round to 1 significant figure ....
multiply by 5 (because 5 moles of carbon dioxide are formed from 1 mole of pentane)	....	round to 1 significant figure ....
multiply by 24 to get dm^3 of carbon dioxide	Answer to 3 significant figures	Answer to 3 significant figures ....

[6]

- We draw bar charts in chemistry to compare values of different physical properties such as melting points, temperatures changes or strength. We can also use them to compare other quantities or numbers.
- The bars are shown as rectangles of the same width which can be vertical or horizontal.
- The bars can be separated from each other or join each other.

The diagram shows two bar charts representing the densities of different elements.

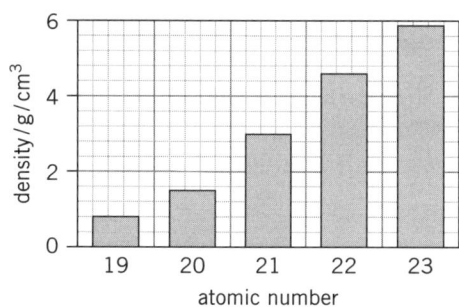 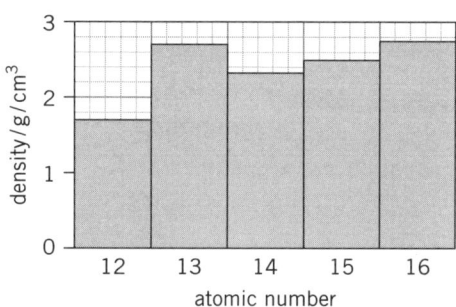

1. The table shows the temperature change when equal amounts of different chlorides **A** to **E** dissolve in the same volume of water.

chloride	A	B	C	D	E
temperature change / °C	+2.5	+4.6	+3.2	−1.5	−0.4

Draw a bar chart to show these results.

- Make sure that you use the full range of the graph paper and label the axes.
- Make sure that you take the negative values into account: The 0 on the y-axis needs to be a little way up the graph paper.

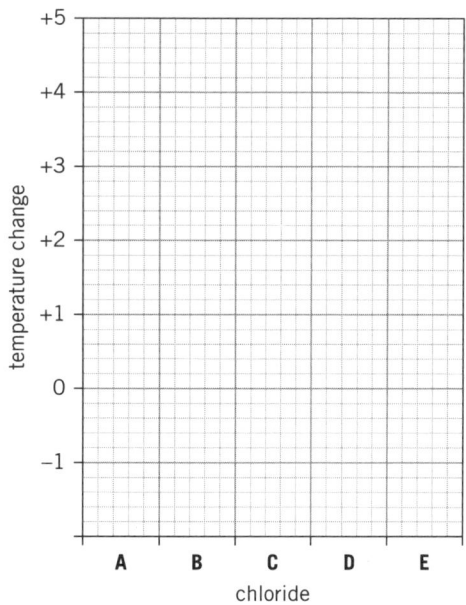

[4]

1. When drawing graphs remember that:
   - The axes should be fully labelled and include units.
   - Use as much of the graph paper as possible.
   - Use an × to plot the points rather than + or • which cannot be so easily seen.
   - If it looks like the line will form a curve, do not use a ruler to join the points to each other.
   - Draw the line of best fit (with equal numbers of points each side of the line if necessary).
   - Ignore any points which do not fit in with the general trend of the line (anomalous points).

   What is wrong with each of these graphs?

   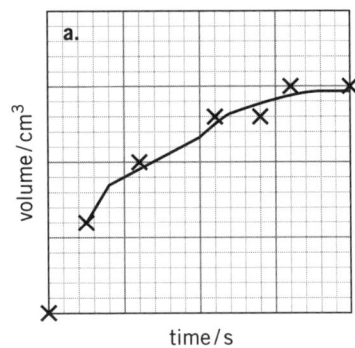   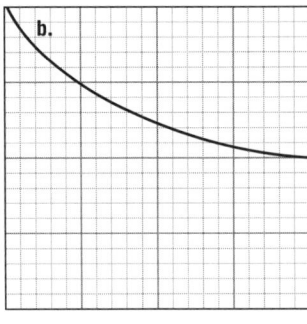   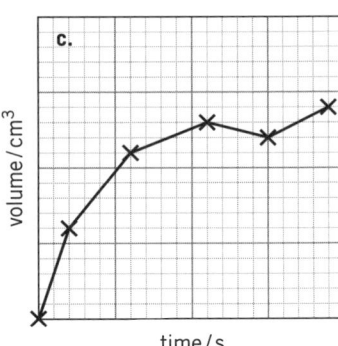

   a. ....

   ....

   .... [3]

   b. ....

   ....

   .... [3]

   c. ....

   .... [2]

2. Extrapolation and Interpolation.

   The graph on the right shows how to extrapolate and interpolate values. Always make sure that you draw lines as shown to the values that are asked for.

   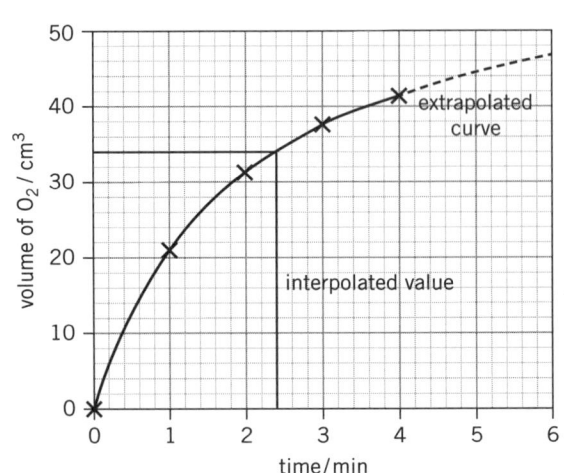

   Deduce the volume of gas released in the first:

   a. 1.4 minutes. .... [1]

   b. 5 minutes .... [1]

1. The table shows how the electrical conductivity of a solution changes as an acid is added to an alkali.

volume of acid / cm³	0	2	3	4	5	6	7	8
conductivity / ohms⁻¹m⁻¹	2.8	1.8	1.3	0.8	0.7	1.0	1.2	1.5

   a. On the grid below plot the points using the data in the table. [3]

   b. Draw two straight lines which intersect at a point. Label this point, P. [2]

   c. Deduce the volume of acid added at point P. .... [1]

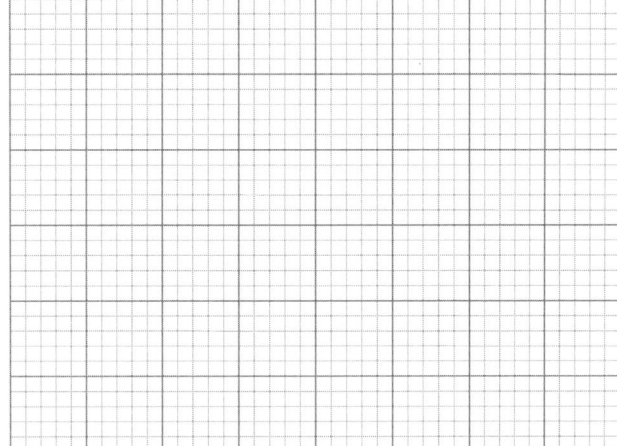

2. The table shows how the volume of gas changes when magnesium reacts with hydrochloric acid. Plot a graph of these results on the grid below. [4]

time / s	0	10	20	30	40	50	60	70
volume of gas / cm³	0	24	39	47	52	54	55	56

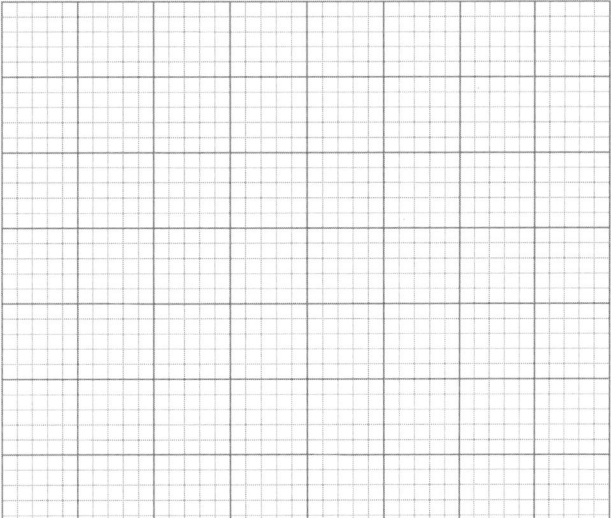

1. Look at the graph below.

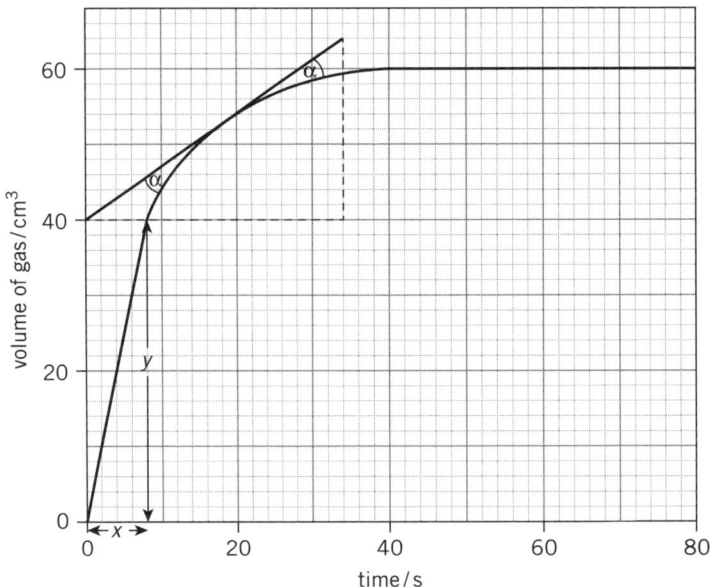

- We can find the initial rate of reaction from a graph by taking the rise, y, over the run, x, of the gradient (slope). This is 40 cm³ ÷ 8 s. So the rate is 5 cm³/s.

- We can find the rate at any other point by drawing a tangent to the curve (see the graph above). Note that the angles α should be equal.

In this case the gradient is $\frac{64 - 40}{34 - 0} = 0.76$ cm³/s

a. The table shows how the mass of a product in a reaction increases with time.

time / s	0	20	40	60	80	100	120	140
mass / g	0	0.08	0.16	0.24	0.30	0.34	0.37	0.40

Plot a graph using these results.

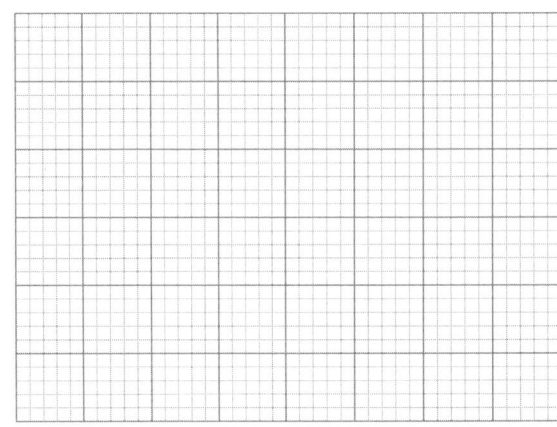

[4]

b. Calculate the initial rate of reaction over the first 40 seconds.

.... [2]

c. Why would the rate calculated over the first 100 seconds be an average rate?

.... [1]

1. Name the most important thing which you wear in the laboratory.

   .... [1]

2. Hydrogen sulfide is a poisonous gas which can be prepared by warming dilute sulfuric acid with iron(II) sulfide. Hydrogen sulfide is heavier than air.

   What precautions would you take when carrying out this experiment in the laboratory?

   .... [2]

3. What is the meaning of each of these hazard symbols?

   A   B   C   D   E

   A ....   B ....   C ....

   D ....   E ....   [5]

4. Describe the hazards associated with each of these chemicals.

   a. Ethanol .... [1]

   b. Concentrated potassium hydroxide .... [1]

   c. Chlorine .... [1]

5. The apparatus below was used to prepare hydrogen chloride gas. Identify and explain three errors.

   sulfuric acid — gas jar

   sulfuric acid

   sodium chloride

   warm

   water

   ....

   ....

   ....

   ....

   .... [6]

1. State five things you need to include when you plan an experiment.

   ....

   ....

   ....

   ....

   .... [5]

   When planning an experiment, you have to think about what things you can change (the variables) and what things you can measure.

   - The variable that you choose the values for is the *independent variable* e.g. 10 s, 20 s, 30 s and so on.
   - The variable that you measure at each of these values is the *dependent variable* e.g.15 cm³ gas at 10 s, 28 cm³ gas at 20 s and so on.
   - All the other values which have to be kept constant to make the experiment a fair test are the *control variables*.

   Identify the three types of variable in the experiments in questions 2 and 3.

2. The effect of temperature on the rate of reaction of hydrochloric acid with calcium carbonate by measuring the volume of carbon dioxide released after 10 s.

   a. Independent variable: .... [1]

   b. Dependent variable: .... [1]

   c. Control variables: ....

      .... [2]

3. The energy released by burning different fuels was compared by measuring the temperature rise of the water in the copper can when 1 g of fuel was burnt.

   a. Independent variable: .... [1]

   b. Dependent variable: .... [1]

   c. Control variables: ....

      .... [2]

1. Accurate results are close to the true value. State two things that will help you get accurate results.

   ....

   .... [2]

2. A student added aqueous 2.0 mol / dm³ hydrochloric acid from a burette to 20 cm³ of a solution of sodium hydroxide in a flask. After each addition of acid, the temperature of the mixture in the flask was measured. Complete the table below by taking the thermometer and burette readings.

burette diagram	total volume of acid added / cm³	thermometer diagram	maximum temperature / °C
0 / 1		25 / 20	
4 / 5		25 / 20	
8 / 9		30 / 25	
12 / 13		30 / 25	

[4]

3. What volumes, in cm³, are shown on the gas syringes **S** and **T**.

   S 10 20 30 40 —part of plunger   **T** 10 20 30 40 —part of plunger

   .... cm³   .... cm³   [2]

1. A flask containing 0.25 mol / dm³ hydrochloric acid was placed on a digital balance and excess calcium carbonate was added. The balance was immediately set to zero and the readings of decrease in mass were taken every 10 seconds. The total decrease in mass at each 10 second interval is given below.

   start 0.0 g, 0.50 g, 0.89 g, 1.24 g, 1.44 g, 1.56 g, 1.68 g, 1.79 g, 1.82 g, 1.82 g

   The results from a repeat experiment are:

   start 0.0 g, 0.72 g, 1.04 g, 1.24 g, 1.44 g, 1.52 g, 1.62 g, 1.74 g, 1.78 g, 1.78 g

   Draw a suitable table to display these results and obtain an average value of the change in mass loss.

   [4]

2. The apparatus shown below can be used to show how reaction rate changes with temperature for the reaction between sodium thiosulfate and hydrochloric acid.

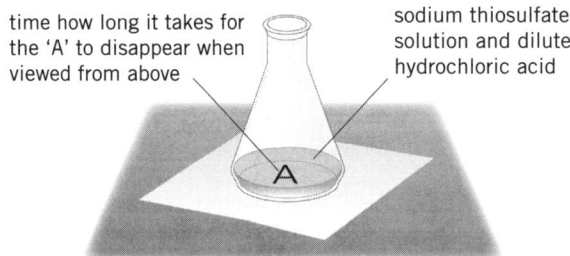

time how long it takes for the 'A' to disappear when viewed from above

sodium thiosulfate solution and dilute hydrochloric acid

   As the reaction proceeds a precipitate of sulfur is made which gradually makes the letter 'A' disappear.

   The rate is proportional to $\dfrac{1}{\text{time taken for the letter 'A' to disappear}}$.

   Draw the headings of a suitable table to enable you to record the results of the experiment at different temperatures.

   [2]

- Numerical data with dependent and independent variables which are both numbers are best displayed on a line graph.
- The independent variable should be the *x*-axis (horizontal axis) of the graph.
- The dependent variable should be the *y*-axis (vertical axis) of the graph.

1. Identify the independent variable and the dependent variable in these two cases.

   a. The electrical conductivity of a solution is measured at 2 minute intervals.

   Independent variable ....

   Dependent variable .... [2]

   b. The rate of reaction of hydrochloric acid with sodium thiosulfate is studied using four different concentrations of acid. At each concentration, the time taken for a precipitate to make a letter M 'disappear' is recorded.

   Independent variable ....

   Dependent variable .... [2]

2. The rate of hydrolysis of compound **A** can be deduced from measurements of electrical conductivity of the reaction mixture. The results are shown in the table.

initial concentration of A in mol / dm³	relative electrical conductivity					
	0 min	1 min	2 min	3 min	3.5 min	4 min
1.4	0	2.5	5.0	7.5	9.2	10.0
2.2	0	3.8	7.5	11.2	13.0	15.0
3.2	0	6.0	12.0	16.0	21.1	24.0
4.0	0	7.5	15.0	22.5	26.0	30.0

   a. On the grid below plot a graph of these results.

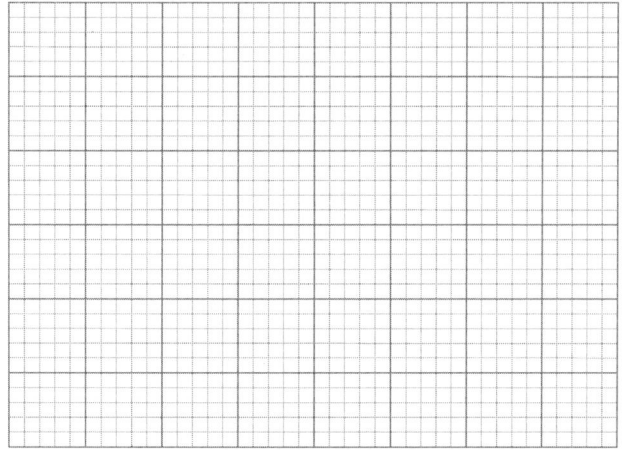

[5]

   b. Which point is anomalous? Explain why.

   ....

   .... [2]

When you have recorded the data, you need to display it in the best way.

- First order your data in a table (see Unit 24.4).
- Then decide what the best way of displaying it, using graphs or charts.
- Numerical data with dependent and independent variables which are both numbers are best displayed on a line graph (see Unit 24.5).
- Data in which one of the variables is not a number is best displayed as a bar chart.
- Sometimes conclusions have to be reached from the data in the table.

1. Hydrogen peroxide, H_2O_2, reacts with acidified iodide ions, I^-. The table shows the relative rate of reaction using different concentrations of hydrogen peroxide and iodide ions (Experiments **A** to **E**).

	concentration of H_2O_2 mol / dm³	concentration of I^- mol / dm³	relative rate of reaction
A	0.01	0.01	1.75
B	0.02	0.01	3.50
C	0.02	0.02	7.00
D	0.02	0.03	10.50
E	0.03	0.01	5.25

a. i. Which three experiments would you compare if you wanted to find how the concentration of hydrogen peroxide affects the rate of reaction?

.... [1]

ii. How does the concentration of hydrogen peroxide affect the rate of the reaction?

.... [2]

b. i. Which three experiments would you compare if you wanted to find how the concentration of iodide ions affects the rate of reaction?

.... [1]

ii. How does the concentration of hydrogen peroxide affect the rate of the reaction?

.... [2]

c. Why is the experiment using three different concentrations of hydrogen peroxide and three different concentrations of iodide ions the minimum amount of data that is needed to draw the correct conclusions about the effect of the concentrations on rate of reaction.

....

.... [2]

2. Look at the graph you plotted in Unit 24.5. How could you use this graph to show that the rate is directly proportional to the initial concentration of **A**?

....

....

.... [3]

136

Conclusions

- Make sure that you describe the patterns in the data accurately, stating the direction of **change** in both the dependent and independent variable, e.g. the rate increases as the concentration increases. (The statement 'Rate increases with concentration' is less accurate because we do not know if concentration is increasing or decreasing.)
- Describe graphs accurately. The line is only proportional if it is a straight upward-sloping line going through the 0–0 point.
- If there is a change in the gradient of a line or curve, make a comment about this e.g. it is a downward sloping curve whose gradient decreases with time.

1. Describe the patterns shown by graphs **A**, **B** and **C** as accurately as possible.

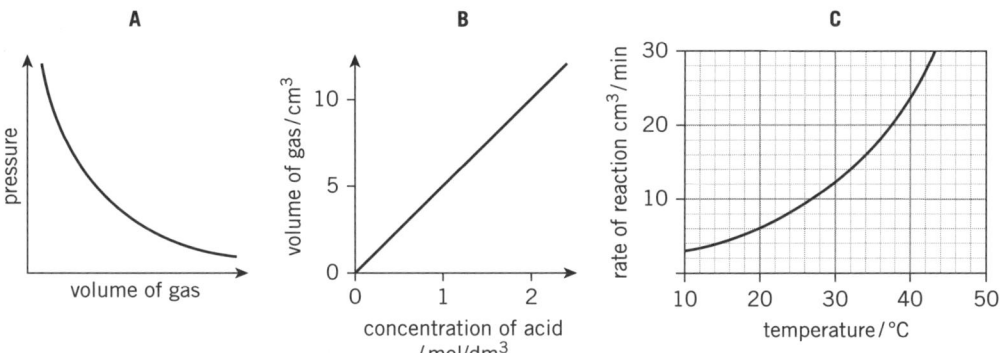

a. Graph **A** ....

   .... [2]

b. Graph **B** ....

   .... [2]

c. Graph **C** ....

   .... [2]

Evaluation

When evaluating an experiment we need to think about:

- The accuracy of the measuring apparatus used. The accuracy of the whole experiment depends on the accuracy of the least accurate measuring instrument.
- The number of times the experiment is repeated. If similar results are obtained when the experiment is repeated several times, we can be sure that the results are correct.
- How easy it is to control the variables. If it is not easy to control a variable, it is not a fair test. For example, if you try to keep a fixed temperature in a beaker of water using a Bunsen burner, you are unlikely to succeed. So use a fixed temperature water bath instead.

Improvement

We can improve our results by:

- Using more accurate measuring apparatus e.g. thermometer with 0.1 °C scale divisions instead of 1 °C scale divisions.
- Repeating the results as many times as possible in the time allowed.
- Making sure that the variables are carefully controlled.

Many salts increase in solubility as the temperature increases. Plan an experiment to see if this is true using the salt potassium chloride as an example. This exercise guides you through the stages in this experiment.

1. Planning:

   a. What apparatus do you need?

   ....

   ....

   .... [5]

   b. What are the independent and dependent variables?

   ....

   .... [2]

   c. What will you keep constant?

   .... [2]

2. Carrying out:
   Describe how you will carry out the experiment. Give possible volumes and masses of the substances used.

   ....

   ....

   ....

   ....

   .... [5]

3. Evaluation:

   a. Explain why it might be difficult to control the temperature.

   ....

   .... [2]

   b. Suggest improvements that you could make to the experiment.

   ....

   .... [2]

   c. Which measurement is most likely to have the greatest percentage error?
      Explain your answer.

   ....

   .... [2]

138

Hydrogen peroxide at a concentration of 2 mol / dm³ can be decomposed by small portions of metal oxides such as manganese(IV) oxide, copper(I) oxide and lead(IV) oxide.

$$2H_2O_2(aq) \rightarrow 2H_2O(l) + O_2(g)$$

Plan an experiment to compare the ability of each of these compounds to decompose hydrogen peroxide.

1. Planning:

   a. What apparatus do you need?

   ....

   ....

   .... [5]

   b. What are the independent and dependent variables?

   ....

   .... [2]

   c. What will you keep constant?

   ....

   .... [3]

2. Carrying out:
   Describe how you will carry out the experiment. Give possible volumes and masses of the substances used.

   ....

   ....

   ....

   ....

   .... [5]

3. Evaluation:

   a. The reaction can be quite fast. Explain any problems associated with this.

   ....

   .... [2]

   b. Suggest improvements that you could make to the experiment.

   ....

   ....

   .... [2]

139

Before the test

- Make sure know how to use apparatus such as a burette, volumetric and dropping pipette.
- Make sure you can take readings correctly from a burette, measuring cylinder and mass balance.
- Make sure that you have plenty of practice at plotting graphs and drawing bar charts.
- Study past test papers carefully to see the types of questions which are asked.

Test questions are of two main types:

- Questions involving a titration, measuring heat changes, following a reaction to deduce rates, electrolysis or determining solubility.
- Questions on qualitative analysis (tests for ions and gases). It helps if you know these thoroughly even if you have access to a data sheet. After all, you have to know them for the theory papers!

What you may be asked to do:

- Read apparatus scales with precision.
- Process data by constructing tables and graphs.
- Interpolate and extrapolate data (especially graphical data).
- Do simple calculations with the aid of a calculator.
- Draw conclusions from the data.
- Evaluate experiments and data.
- Identify sources of error and suggest improvements to an experiment.

During the test

- Read the questions carefully to make sure you know exactly what to do.
- Keep calm and divide your time appropriately between the (two) experiments.
- If you make a mistake in your answer, delete the incorrect answer with a single line and rewrite the answer so that the marker is aware of where it is.
- If you make a practical mistake start the exercise again if you have enough material.
- Record the temperature to nearest °C.
- Use the amount of materials given in the question. For qualitative analysis questions use the minimum amount of material for the experiment. About 1 cm depth of solution is usually enough.
- Don't throw any substance away (you may need it if you make a mistake or need it for an additional test).
- Label your test tubes. This is especially important when you need a solution for an additional test or there are several different colourless solutions.
- Don't use the same dropping pipette for different solutions. If necessary, wash the pipette out with distilled water and make sure that it is fairly dry before using it for a different solution.
- When adding a solution from a dropping pipette to another solution in a test tube, add the solution slowly so that you can observe any gradual changes.

Writing your observations

- Make sure that you write your observations as you go along, not after doing several operations.
- Make sure that you include the colour changes of the solutions if any.
- Make sure that the states are included e.g. precipitate formed / bubbles of gas.

Introduction

This could be done at home or in the school laboratory.

- Hard water does not lather well with soap.
- Soft water lathers well with soap.
- Permanent hardness in water cannot be removed by boiling.
- Temporary hardness in water can be removed by boiling.

Purpose of the experiments

To find the volume of soap solution (or washing up liquid) needed to form a permanent lather with different samples of water.

Sources of water

- Distilled water
- Temporary hard water (bubble carbon dioxide through limewater until the white precipitate has disappeared)
- Permanent hard water (add some hydrated calcium sulfate to distilled water then filter)
- Natural sources of water e.g. tap water, rainwater, seawater.

Carrying out the experiments

1. There are two ways in which the experiment can be carried out.
   a. Add the water sample to a flask and then see how many drops of soap solution (added from a burette or pipette are needs to form a lather that does not disappear on shaking then leaving for a minute.
   b. Adding a certain number of drops of soap solution to a tube of water, shaking and measuring the height of the lather formed.
2. Make a list of all the equipment that you need including safety equipment/clothing.
3. What do you need to vary and what do you need to keep constant?

Analysing the results

- Draw a table of results for the number of drops (or height of lather) with different types of water including boiled hard water (temporary and permanent).
- Repeat your experiments to get consistent results.
- Suggest how you could improve your experiments.

Conclusions

1. Which samples of water are hard and which are soft?
2. Classify tap water, rainwater, seawater and other sources of water you have analysed as hard or soft.
3. Which samples contain temporary hardness and which contain permanent hardness?

Introduction

This could be done at home or in the school laboratory.

- The labels on packets of food usually state the amount of energy they contain in kilojoules or kilocalories. Make a list of these values for the foods you choose.
- When dry foods are burnt, they release energy. The reaction is exothermic.
- The carbohydrate, fats and proteins burn to form carbon dioxide and water.

Purpose of the experiments

To compare the energy released by different foods.

Sources of foods

- The foods should be dry. You can dry wet foods in an oven but don't let them char (go black).
- Make a list of the energy values for the foods you choose by looking on the sides of the packets.
- Crisps, bread, nuts, rice are good sources.
- Dried meats, beans and cheese could also be used.
- It is possible to burn cooking oils if you use a cotton or string wick.

Carrying out the experiments

1. There are two ways in which the experiment can be carried out.

   a. Burning different foods of known mass on the end of a large needle. The burning foods heat a known volume of water in a test tube or beaker.

   b. Burning the food on a tin lid beneath a beaker or tin of water. This is more useful for fats and oils.

2. Make a list of all the equipment that you need including safety equipment/clothing.

3. What do you need to vary and what do you need to keep constant?

4. You could also investigate the relationship between the mass of a particular food burnt and the temperature rise.

Analysing the results

- Draw a table of results for the temperature rise on burning a known amount of food material using a fixed volume of water.
- Repeat your experiments to get consistent results.
- Suggest how you could improve your experiments.

Conclusions

1. Calculate the energy released in kJ per gram of food by using the relationship
   energy released (Joules) = mass of water (g) × 4.18 × temperature rise (°C).

2. Which foods released the most energy per gram?

3. Compare the energy values you obtained with the energy values on the labels of the foods you used. Were they in the same order of energy as the results of your experiments? If not, suggest why not.

4. Suggest reasons why your experiment may not be a fair test.

Introduction

This is best done in the laboratory.

- Many compounds are much more soluble in water at higher temperatures than lower temperatures. Others do not show much difference in solubility as the temperature increases.

Purpose of the experiments

To find how the solubility of different compounds changes with temperature.

Suggested compounds to use

Potassium nitrate, sodium nitrate, potassium chloride and sodium chloride

Carrying out the experiments

1. **a.** Heat some water (4 or 5 cm³) with the solute until the solute dissolves.

   b. The solution is then cooled using the apparatus shown below until crystallisation occurs. The temperature of crystallisation is recorded.

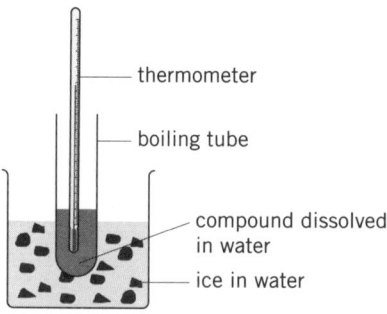

   c. Then add more water (not more than 2 cm³) to the solution in the boiling tube and heat to dissolve. The temperature when crystals appear is recorded.

   d. Repeat step **c.** several times.

2. Make a list of all the equipment that you need including safety equipment/clothing.

3. What do you need to vary and what do you need to keep constant?

Analysing the results

- For each salt draw a table of results for the temperature at which crystallisation occurs.
- Repeat your experiments to get consistent results.
- Suggest how you could improve your experiments.

Conclusions

1. Which compounds show the greatest difference in crystallisation temperature?
2. Compare your results with the tables showing the solubility of each of these compounds at different temperatures.
3. Suggest reasons why your experiment may not be a fair test.

Introduction

This should be done in the laboratory.

- When carbonates react with acids carbon dioxide is released.
- The mass of the reaction mixture decreases as the reaction proceeds.
- The diagram below shows some of the apparatus that can be used to follow the rate of this reaction.

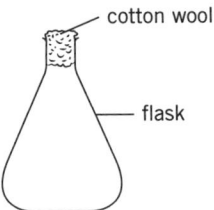

cotton wool

flask

Purpose of the experiments

To find the percentage of carbon dioxide and hence the percentage by mass of carbon in different carbonates.

Suggested carbonates to use

Sodium carbonate, sodium hydrogencarbonate, calcium carbonate, copper(II) carbonate, barium carbonate.

Carrying out the experiments

1. The hydrochloric acid needs to be in excess. Why?
2. Deduce the amounts of hydrochloric acid and calcium carbonate you need.
3. Make a list of all the equipment that you need including safety equipment/clothing.
4. What do you need to vary and what do you need to keep constant?

Analysing the results

- Draw up a table of results for the volume of carbon dioxide given off for each carbonate.
- Repeat your experiments to get consistent results.
- Suggest how you could improve your experiments.
- Calculate the mass of carbon released in each experiment.
- Calculate the percentage by mass of the carbon in the carbonate.

Conclusions

1. Put the carbonates in order of their percentage composition carbon by mass.
2. The CO_3^{2-} is common to all the carbonates. So why are the percentage compositions different?
3. Suggest reasons why your experiment may not be a fair test.

1. Phosphorus is an element in Group V of the Periodic Table.

   a. Deduce the electronic configuration of an atom of phosphorus.

   .... [1]

   b. An isotope of phosphorus has 15 protons and 31 nucleons.

   Deduce the number of neutrons in this isotope of phosphorus.

   .... [1]

   c. Phosphorus has a simple molecular structure.

   Describe two physical properties of phosphorus.

   ....

   .... [2]

   d. Phosphorus burns in excess oxygen to form an oxide with the formula P_2O_5.

   Write a balanced equation for this reaction.

   .... [2]

   e. P_2O_5 reacts with sodium hydroxide to form sodium phosphate, Na_3PO_4.

   Deduce the formula of the phosphate ion.

   .... [1]

   f. Phosphate ions are present in many fertilisers.

   Name another anion that is present in most fertilisers.

   .... [1]

   g. Explain why farmers spread fertilisers on the soil where crop plants are grown.

   ....

   .... [2]

   h. Draw the electronic configuration of phosphine, PH_3. Show only the outer shell electrons.

   [2]

   Total = 12

2. The structure of allyl alcohol is shown below.

$$CH_2=CH-CH_2-OH$$

   a. What feature of allyl alcohol shows that it is an unsaturated compound?

.... [1]

   b. Describe a test for an unsaturated compound.

Test ....

Result .... [2]

   c. Allyl alcohol can be reduced by hydrogen in a similar way to ethene.

      i. Explain the term reduction in terms of electron transfer.

.... [1]

     ii. State the conditions needed for this reduction.

....

.... [3]

    iii. Give the displayed formula of the compound formed by this reduction. Show all atoms and all bonds.

[2]

   d. Compounds with structures similar to allyl alcohol are found in green onion leaves.

      i. Suggest how you could make a solution of the green pigments from onion leaves.

....

.... [2]

     ii. Several green pigments are present in onion leaves. State the name of the method you would use to separate these pigments from each other.

.... [1]

Total = 12

146

3. The structure of caesium chloride is shown below.

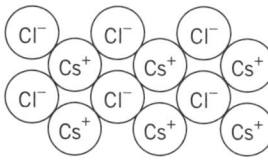

a. Deduce the empirical formula of caesium chloride.

.... [1]

b. Explain in terms of structure and bonding why caesium chloride has a high melting point.

....

.... [2]

c. Explain why aqueous caesium chloride conducts electricity.

.... [1]

d. Molten caesium chloride is electrolysed using graphite electrodes.

   i. Give two reasons why graphite electrodes are used.

   ....

   .... [2]

   ii. Write ionic half equations (ion electron equations) for the reactions at:

   the anode ....

   the cathode .... [3]

e. Caesium chloride is formed when caesium burns in chlorine.

$$2Cs \ + \ Cl_2 \ \rightarrow \ 2CsCl$$

When 5.32 g of caesium are burnt in excess chlorine, 6.4 g of caesium chloride are formed. Calculate the percentage yield of caesium chloride.

[3]

Total = 12

4. A student investigated the reaction at r.t.p. between 0.05 g magnesium ribbon and excess hydrochloric acid of concentration 2.0 mol/dm³.

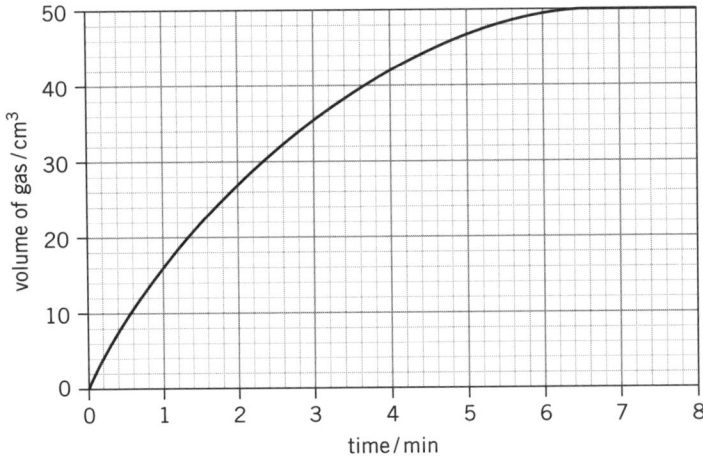

a. Deduce the time when the reaction was just complete.

.... [1]

b. i. Deduce the volume of hydrogen released during the first minute of the reaction.

.... [1]

ii. Deduce the average rate of reaction during the first two minutes.

.... [1]

c. The experiment was repeated at r.t.p. using hydrochloric acid of concentration 2.5 mol/dm³.
On the grid above draw a line to show how the volume of hydrogen released changes with time. [2]

d. State and explain using the collision theory how increasing the concentration of acid affects the rate of reaction.

....

....

.... [2]

e. The experiment was repeated using 2 mol/dm³ hydrochloric acid and 0.05 g magnesium powder.
Would the reaction be faster or slower? Explain your answer.

....

.... [2]

Total = 9

5. When 1 mole of calcium carbonate is heated, 1 mole of calcium oxide and 1 mole of carbon dioxide are formed.

   a. What type of reaction is this? Put a ring around **two** of the words below.

   addition   catalysed   decomposition   endothermic

        exothermic   oxidation   reduction   [2]

   b. Describe a test for carbon dioxide.

   Test ....

   Result .... [2]

   c. The table shows the mass of calcium carbonate converted to products in 5 minutes at different temperatures. The same mass of calcium carbonate was used in each experiment.

temperature / °C	500	700	800	900	950	1000
mass converted	0.0	0.4	1.8	3.5	3.7	3.8

      i. On the grid below draw a graph of these results.

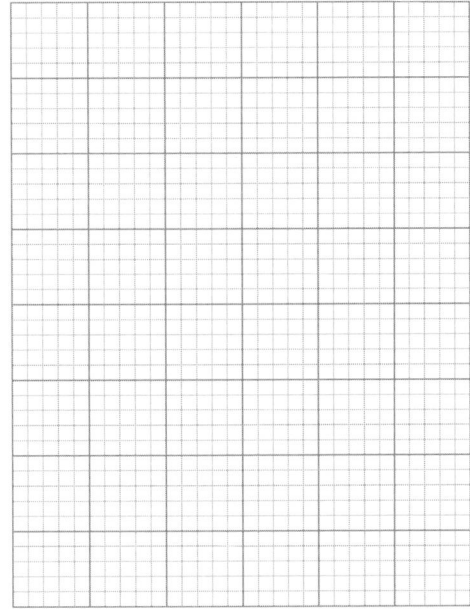

[3]

      ii. Use your graph to help you calculate the volume of carbon dioxide formed when calcium carbonate is heated for 5 minutes at 850 °C.

[3]

Total = 10

6. The diagram shows the preparation of ammonia by heating ammonium sulfate with concentrated sodium hydroxide.

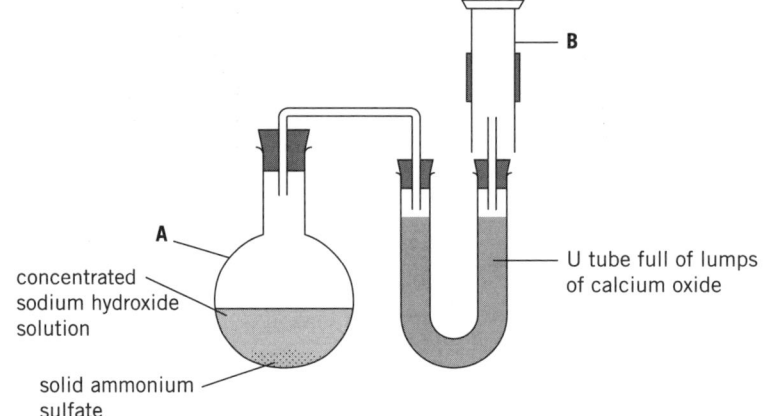

a. i. On the diagram above, show where heat is applied. [1]

ii. State the names of the pieces of apparatus labelled **A** and **B**.

A ....

B .... [2]

iii. What is the purpose of the calcium oxide?

.... [1]

iv. Explain how you can show when **B** is full of ammonia.

....

.... [2]

b. Complete the equation for the reaction.

$(NH_4)_2SO_4$ + ... → .... + ... + H_2O [2]

c. Hydrazine, $H_2N–NH_2$, like ammonia, contains hydrogen and nitrogen.

Draw the electronic configuration of a molecule of hydrazine. Show only the electrons in the outer shells.

[2]

Total = 10

7. The structure of lactic acid is shown below.

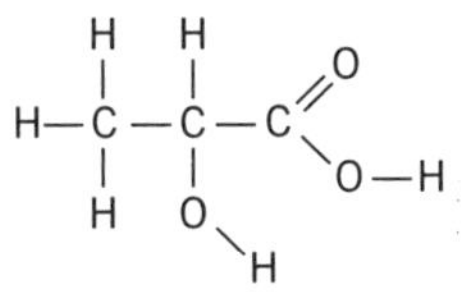

a. On the structure above put a ring around the alcohol functional group. [1]

b. Lactic acid can be made by fermenting the sugar lactose.

   i. State the three different types of atom present in sugars.

   .... [2]

   ii. Give the name of another compound which can be made by fermentation.

   .... [1]

c. Calcium carbonate neutralises lactic acid. Complete the word equation for this reaction.

   lactic acid   +   calcium   →   calcium   + ...   + ...
   carbonate   lactate   .... [2]

d. Calcium lactate is insoluble in water. Suggest how you could separate calcium lactate from a mixture of calcium lactate and aqueous salts.

   .... [1]

e. The simplified structure of the polymer of lactic acid is shown below.

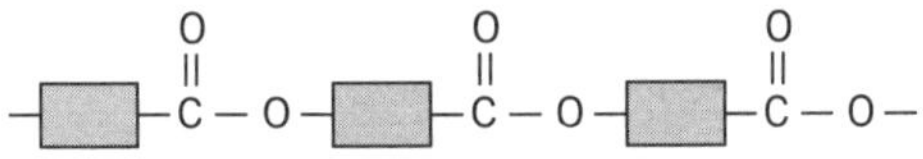

   i. State the name of the linkage group.

   .... [1]

   ii. Explain why this is not an example of addition polymerisation.

   ....

   .... [2]

f. Lactic acid is oxidised to ethanoic acid by acidified potassium manganate(VII). What colour change would you observe when excess lactic acid is added to acidified potassium manganate(VII)?

   .... to .... [2]

   Total = 12

8. The table shows some physical properties of four noble gases.

gas	melting point / °C	boiling point / °C	density at r.t.p. in g / dm^3	atomic radius / nm
helium	−272	−269	0.18	0.050
neon	−248	−246	0.90	0.065
argon	−189	−186	1.78	
krypton	−157	−152	3.74	0.110

a. i. The density of air is 1.20 g / dm³. Which of these gases could be used to fill a toy balloon to float in air?

.... [1]

ii. Deduce the atomic radius of argon.

.... [1]

iii. What is the state of krypton at −118 °C? Explain your answer.

.... [2]

iv. Describe the trend in boiling point down the Group.

.... [1]

b. Xenon tetrafluoride, XeF_4, reacts with potassium iodide.

$$XeF_4 \ + \ 4KI \ \rightarrow \ Xe \ + \ 2I_2 \ + \ 4KF$$

i. Potassium salts are colourless. What is the final colour of the reaction mixture?

.... [1]

ii. Use the equation to explain how you know that potassium iodide is acting as a reducing agent in this reaction.

....

.... [2]

iii. Calculate the maximum volume of xenon formed when 8.28 g of xenon tetrafluoride reacts with excess potassium iodide.

[3]

Total = 11

9. Nitrogen dioxide, NO_2, is a brown gas which pollutes the atmosphere.

    a. **i.** Give one source of nitrogen dioxide in the atmosphere.

         .... [1]

      ii. Describe one effect of nitrogen dioxide on the environment.

         .... [1]

      iii. Nitrogen dioxide is a gas at r.t.p. Describe the proximity (closeness) and motion of the particles in nitrogen dioxide at r.t.p.

         ....

         .... [2]

    b. The colourless gas dinitrogen tetroxide, N_2O_4, forms an equilibrium mixture with nitrogen dioxide.

$$N_2O_4(g) \rightleftharpoons 2NO_2(g)$$

      i. Describe and explain what you would observe when the pressure on this equilibrium mixture is increased.

         ....

         ....

         .... [3]

      ii. Calculate the relative molecular mass of

         nitrogen dioxide ....

         dinitrogen tetroxide .... [2]

      iii. At 55 °C the average relative molecular mass of the equilibrium mixture is 61.0 but at 140 °C, the average relative molecular mass is 46.0.

         Explain how this shows that the reaction is endothermic.

         ....

         ....

         .... [3]

    c. At temperatures above 150°C nitrogen dioxide decomposes to nitrogen(II) oxide and oxygen. Write a symbol equation for this reaction.

    .... [2]

                                                       Total = 14

10. 25 cm³ of aqueous potassium hydroxide was placed in a flask. A few drops of an acid–base indicator were then added. The solution was neutralised by 12.5 cm³ of 0.2 mol / dm³ sulfuric acid added from a burette.

$$2KOH(aq) \quad + \quad H_2SO_4(aq) \quad \rightarrow \quad Na_2SO_4(aq) \quad + \quad 2H_2O(l)$$

a. Suggest a suitable indicator that could be used in this reaction.

.... [1]

b. Give the name of the salt formed in this reaction.

.... [1]

c. Calculate:

  i. The number of moles of sulfuric acid added from the burette.

  [1]

  ii. The number of moles of potassium hydroxide in the flask.

  .... [1]

  iii. The concentration of potassium hydroxide in the flask in mol / dm³.

  [1]

d. Write the simplest ionic equation for this reaction.

.... [1]

e. Sulfuric acid catalyses the reaction between ethanol and butanoic acid.

Draw the displayed formula for the ester formed in this reaction, showing all atoms and all bonds.

[2]

Total = 8

Acid: A proton donor.

Acidic oxide: An oxide which reacts with alkalis to form a salt and water.

Acid rain: Rain which has a pH below 5.6 due to the reaction of rainwater with acidic gases.

Activation energy: The minimum amount of energy particles must have to react when they collide.

Addition polymerisation: Polymerisation of monomers containing a C=C double bond to form a polymer and no other compound is formed.

Addition reaction: A reaction in which a single product is formed from two or more reactant molecules and no other product is made.

Alcohols: Organic compounds with branched or unbranched chains containing the –OH functional group.

Alkali: A base which is soluble in water.

Alkanes: Saturated hydrocarbons with the general formula C_nH_{2n+2}.

Alkenes: Hydrocarbons containing at least one C=C double bond.

Alloy: A mixture of a metal with another element.

Amphoteric oxide: An oxide which reacts with both acids and alkalis.

Anions: Negative ions.

Anode: The positive electrode.

Atom: The smallest particle that cannot be broken down by chemical means.

Atomic number: The number of protons in the nucleus of an atom.

Avogadro constant: The number of particles in a mole of defined particles (atoms, ions or molecules).

Base: A proton acceptor.

Basic oxide: An oxide which reacts with acids to form a salt and water.

Carboxylic acids: A homologous series of organic compounds with the –COOH group.

Catalyst: A substance which speeds up a chemical reaction but remains unchanged at the end of the reaction.

Catalytic converter: Part added to vehicle to reduce the emissions of carbon monoxide and nitrogen oxides from exhausts of petrol engines.

Cathode: The negative electrode.

Cations: Positive ions.

Collision theory: The theory that moving particles react when they collide with sufficient energy and in the correct orientation.

Compound: A substance made up of two or more different atoms (or ions) joined together by bonds.

Condensation polymerisation: Polymerisation occurring when two types of monomer bond together with the elimination of small molecule.

Condensing: The change of state from gas to liquid.

Conductors (electrical): Substances which have a low resistance to the passage of electricity.

Corrosion: The gradual reaction and 'eating away' of a metal inwards from its surface caused by another substance.

Covalent bond: A shared pair of electrons.

Cracking: The decomposition of larger alkane molecules into a mixture of smaller alkanes and alkenes.

Delocalised electrons: Electrons which are not associated with any particular atom.

Diatomic: Molecules containing two atoms.

Diffusion: The spreading movement of one substance through another due to the random movement of the particles.

Displayed formula: Shows how the atoms and bonds in a compound are arranged.

Dot and cross diagram: A diagram showing the electronic configuration of atoms ions or molecules.

Double bond: Two covalent bonds between the same two atoms.

Ductile: Can be drawn into wires.

Electrochemical series: The order of reactivity of metals, with the most reactive at the top.

Electrodes: Rods which conduct electric current to and from an electrolyte.

Electrolysis: The decomposition of a compound when molten or in aqueous solution by an electric current.

Electrolyte: A molten ionic compound or a solution containing ions which conducts electricity.

Electron: The negatively charged particles outside the nucleus of an atom.

Electron shells: Spherical areas surrounding the nucleus which contain one or more electrons.

Electroplating: Coating of the surface of one metal with a layer of another, usually less reactive, metal.

Element: A substance made up of only one type of atom which cannot be broken down into anything simpler by chemical reactions.

Empirical formula: Shows the simplest whole number ratio of atoms or ions in a compound.

Endothermic reaction: A reaction which absorbs energy from the surroundings.

Enthalpy change: The heat energy exchanged between a chemical reaction and its surroundings at constant pressure.

Enzymes: Biological catalysts.

Ester: A compound with the formula R–COO-R′ formed by the reaction of an alcohol with an alkanoic acid.

Esterification: Making an ester by the reaction of an alcohol with an alkanoic acid.

Evaporation: The change of state from liquid to vapour which takes place below the boiling point of a liquid.

Exothermic reaction: A reaction which releases energy to the surroundings.

Fermentation: The breakdown of organic materials by microorganisms with effervescence and the release of heat energy.

Filtrate: The solution passing through a filter paper when a mixture of solid and solution are filtered.

Flue gas desulfurisation: Removal of sulfur dioxide in industry arising from burning fossil fuels containing sulfur.

Fraction: A product of petroleum distillation which is a mixture of hydrocarbons having a limited range of molar masses and boiling points.

Fractional distillation: A method used to separate two or more liquids with different boiling points from each other using a distillation column.

Freezing: The change of state from liquid to solid.

Functional group: A group which is characteristic of a given homologous series.

General formula: A formula which can be applied to all members of a given homologous series.

Giant molecular structure: A structure having a three-dimensional network of covalent bonds.

Global warming: The heating of the atmosphere caused by absorption of infra-red radiation by greenhouse gases.

Greenhouse gases: Gases which are good absorbers of infrared radiation and cause global warming.

Group: A vertical column in the Periodic Table.

Half equations: Equations showing the oxidation and reduction reactions separately.

Halogens: The elements in Group VII.

Homologous series: A group of compounds with the same general formula and the same functional group.

Hydrocarbons: Compounds containing only carbon and hydrogen atoms.

Hydrogenation: A reaction involving the addition of hydrogen to a compound.

Hydrolysis: The breakdown of a compound by water often catalysed by acids or alkalis.

Incomplete combustion: Combustion when air or oxygen is limiting.

Indicator: A compound or mixture of coloured compounds which changes colour over a specific pH.

Insulators: Non-conductors.

Ion: An atom or group of atoms with either a positive or negative charge.

Ionic bond: The strong force of attraction between oppositely charged ions.

Ionic equation: A symbol equation that shows only those ions and molecules which take part in a reaction.

Isotopes: Atoms of elements with the same number of protons but different numbers of neutrons.

Kinetic particle theory: The idea that particles are in constant motion.

Lustrous: Having a shiny surface.

Macromolecules: Very large molecules made up of repeating units.

Malleable: Can be shaped by hitting.

Mass number: The number of protons + the number of neutrons in an atom.

Melting: The change of state from solid to liquid.

Metallic bond: A bond formed by the attractive forces between the delocalised electrons and the positive ions.

Metallic conduction: The movement of mobile electrons through the metal lattice when a voltage is applied.

Mixture: This consists of two or more elements or compounds which are not chemically bonded together.

Molar concentration: The number of moles of solute dissolved in a solvent to make 1 dm^3 of a solution.

Molar gas volume: The volume of a mole of gas at r.t.p. or s.t.p.

Molar mass: The mass of a substance in moles.

Mole: The amount of substance that contains 6.02×10^{23} defined particles (atoms, ions or molecules).

Molecular equation: A full symbol equation.

Molecular formula: Shows the number of atoms of each particular element in one molecule of a compound.

Molecule: A particle containing two or more atoms. The atoms can be the same or different.

Monomers: The small molecules which react and bond together to form a polymer.

Neutralisation: The reaction between an acid and a base to form a salt and water.

Neutral oxide: An oxide which does not react with acids or alkalis.

Neutron: The neutral particle in the nucleus of an atom.

Noble gas configuration: Atoms or ions having a complete outer shell of electrons.

Nucleus: A tiny particle in the centre of an atom containing protons and neutrons.

Oxidation: The gain of oxygen or loss of electrons by a substance.

Oxidation number: A number given to each atom or ion in a compound to show the degree of oxidation.

Oxidising agent: A substance which accepts electrons and gets reduced.

Paper chromatography: A method used to separate a mixture of different dissolved substances depending on the solubility of the substances in the solvent and their attraction to paper.

Percentage yield: $\dfrac{\text{actual yield of product}}{\text{theoretical yield of product}} \times 100$

Periodic Table: Arrangement of elements in order of increasing atomic number so that most Groups contain elements with similar properties.

Periodicity: The regular occurrence of similar properties of the elements in the Periodic Table so that some groups have similar properties or a trend in properties.

Period: A horizontal row in the Periodic Table.

Petroleum: A thick liquid mixture of unbranched, branched and ring hydrocarbons extracted from beneath the Earth's surface.

Photochemical reaction: A reaction which depends on the presence of light.

pH scale: A scale of numbers from 0 to 14 used to show how acidic or alkaline a solution is.

Physical properties: Properties which do not generally depend on the amount of substance present.

Pollution: Contaminating materials introduced into the natural environment (earth, air or water).

Polyamide: Condensation polymer containing –NH–CO– linkages.

Polyester: Condensation polymer containing –COO– linkages.

Polymerisation: The conversion of monomers to polymers.

Polymers: Macromolecules made up by linking at least 50 monomers.

Precipitate: The solid obtained in a precipitation reaction.

Precipitation reaction: A reaction in which a solid is obtained when solutions of two soluble compounds are mixed.

Proton: The positively charged particles in the nucleus of an atom.

Rate of reaction: The change in concentration of a reactant or product with time at a stated temperature.

Reaction pathway diagram: A diagram showing the enthalpy change from reactants to products for exothermic or endothermic reactions (the activation energy can also be shown).

Redox (reaction): A reaction where there is simultaneous oxidation and reduction.

Reducing agent: A substance which loses electrons and gets oxidised.

Reduction: The loss of oxygen or gain of electrons by a substance.

Relative atomic mass: The weighted average mass of the isotopes of an element compared to 1/12th the mass of an atom of carbon–12.

Relative formula mass: The relative mass of one formula unit of a compound compared to 1/12th the mass of an atom of carbon–12.

Relative molecular mass: The relative mass of one molecule of a compound compared to 1/12th the mass of an atom of carbon–12.

Repeat unit: A regularly repeating part of a polymer.

Residue: The solid remaining on the filter paper when a mixture of solid and solution are filtered.

r.t.p.: Room temperature and pressure. (20 °C and 1 atmosphere pressure).

Rusting: Corrosion of iron and iron alloys caused by the presence of both water and oxygen.

Salt: A compound formed when the hydrogen in an acid is replaced by a metal or ammonium ion.

Saturated compounds: Organic compounds with only single bonds.

Separating funnel: Piece of apparatus used to separate immiscible liquids which have different densities.

Simple distillation: The separation of a liquid from a solid which involves the processes of boiling and condensation using a condenser.

Solubility: The number of grams of solute needed to form a saturated solution per 100 grams of solvent used.

Solution: A uniform mixture of two or more substances.

Solute: A substance that is dissolved in a solvent.

Solvent: A substance that dissolves a solute.

Sonorous: Rings when hit with a hard object.

Spectator ions: Ions which do not take part in a reaction.

Standard concentration: A concentration of 1 mole of substance in 1 dm^3 of solution under standard conditions.

State symbols: Letters put after a chemical formula showing whether it is a solid, liquid, gas or aqueous solution.

Strong acid: An acid which ionises completely in solution.

Strong base: A base which ionises completely in solution.

Structural formula: Shows the way the atoms are arranged in a molecule with or without showing the bonds.

Structural isomers: Compounds with the same molecular formula but different structural formulae.

Substitution reaction: A reaction in which one atom or group of atoms replaces another.

Thermal decomposition: The breakdown of a compound when heated.

Titration: A method used to determine the amount of substance present in a given volume of solution of acid or alkali.

Titre: The final burette reading minus the initial burette reading in a titration.

Triple bond: Three covalent bonds between the same two atoms.

Unbranched hydrocarbons: Hydrocarbons with carbon atoms linked in a chain without alkyl side groups.

Unsaturated compounds: Organic compounds containing double or triple carbon–carbon bonds (in addition to single bonds).

Volatile: Easily evaporated at room temperature.

Weak acid: An acid which only partially dissociates / ionises in solution.

Weak base: A base which only partially dissociates / ionises in solution.

Unit 1.1

LL In solids the particles are **regularly** arranged and close to each other. The particles only **vibrate**. They do not **move** from place to place. In liquids, the particles are not arranged in a **fixed** pattern and are **close** together. The particles move by **sliding** over each other. In gases, the particles are **far** apart and are able to move **everywhere** rapidly.
(1 mark for each word in the correct place) [8]

1. Box B: (solid): particles touching each other [1]
   particles arranged regularly / in more than 1 regular row [1]
   Box C: (liquid): particles touching each other [1]
   particles arranged irregularly / not in rows [1]

2. A: liquid; B: gas; C: solid; D: liquid [4]

3. a. b.

Axes correctly labelled for (a) AND (b) [1]
(a) downward concave curved line [1]
(b) straight line sloping upwards cutting the y axis or through 0-0 point [1]

4. Particles in gases behave as hard **spheres**. They move in all **directions**. They do not **attract** each other. [3]

Unit 1.2

LL When a solid is heated, the increase in **energy** makes the particles **vibrate** more. The forces of **attraction** between the particles are weakened. At the **melting** point, these forces are **weak** enough for the particles to be able to move and slide over each other. When the liquid is at its **boiling** point, the particles have enough energy to **escape** from the **surface** of the liquid. [8]
(1 mark for each word in the correct place)

1. A: melting / fusion [1]
   B: boiling / evaporation [1]
   C: freezing [1]
   D: condensing [1]

2. a. Diagram showing plunger pushed down so volume smaller
      and the same number of particles randomly arranged [1]
   b. More particles hit the wall [1]
      Every second [1]
      So the force on the wall is greater [1]

3. Particles move faster / particles have more kinetic energy [1]
   Particles hit the walls of the container with more force [1]
   Force per unit area greater at higher temperature / pressure increases [1]

Unit 1.3

LL When a liquid above room temperature cools the **kinetic** energy of the particles **decreases**. The **temperature** of the liquid falls. At the melting point the temperature stays **constant** for a time. This is because thermal energy (heat) is being **released** when a liquid **freezes**. [6]
(1 mark for each word in the correct place)

1. a. Methane [1]
   b. Naphthalene [1]
      Melting point is above room temperature [1]
   c. Ethanol [1]
      Melting point is below room temperature **and** boiling point is above room temperature / room temperature is between melting point and boiling point [1]

2. A: solid **and** liquid [1]
   B: liquid [1]
   C: liquid **and** vapour / liquid **and** gas [1]
   D: vapour / gas [1]

3. When a gas cools the particles have less kinetic energy [1]
   Temperature remains constant at the boiling point [1]
   Because energy is being given out [1]
   To condense gas to liquid [1]

Unit 1.4

LL A: solvent [1]
   B: solute [1]
   C: solution [1]
   D: aqueous [1]
   E: saturated [1]

1. a. white [1] to blue [1]
   b. blue [1] to pink / red [1]

2. a. Has water (of crystallisation) in it [1]
   b. A specific amount of water in the lattice structure of a crystal [1]

3. a. 0.25 dm^3 [1]
   b. 0.15 dm^3 [1]
   c. 0.05 dm^3 [1]
   d. 0.025 dm^3 [1]

4. a. $5 \times 500/1000 = 2.5$ g / dm^3 [1]
   b. $15 \times 400/1000 = 6$ g / dm^3 [1]
   c. $2 \times 40/1000 = 0.08$ g / dm^3 [1]
   d. $0.5 \times 25/1000 = 0.0125$ g / dm^3 [1]

5. moles [1]

Unit 1.5

LL In liquids and gases, the **particles** are constantly moving and **changing** direction when they **hit** other particles. We say that they move **randomly**. Diffusion is the random **movement** of particles in any direction so that they get **mixed** up. Diffusion in **gases** is faster than in **liquids** because the particles move faster in gases. [8]
(1 mark for each word in the correct place)

1. a. Diffusion [1]
   b. Particles (of dye and water) move randomly / move in any direction [1]
      Dye particles spread out [1]
      Overall movement of the dye is from area of high concentration (of the dye particles) to lower concentration (of dye particles) [1]

2. White solid forms where hydrogen chloride reacts with ammonia [1]
   Hydrogen chloride has a higher relative molecular mass than ammonia [1]
   So rate of diffusion of hydrogen chloride less than that of ammonia [1]

3. Chlorine [1] it has the highest relative molecular mass [1]

Unit 2.1

LL Test tube; Burette; Pipette; Flask; Beaker; Syringe; Timer; Balance [8]
(1 mark each)

1. a. A: burette [1]
      B: volumetric flask [1]
      C: measuring cyclinder [1]
      D: volumetric pipette [1]
   b. i. B [1]
      ii. D [1]

2. B [1]

Unit 2.2

LL The method of separating a **mixture** of coloured substances using **filter** paper is called chromatography. The colours **separate** if they have different **solubilities** in the solvent and different degrees of **attraction** for the filter paper. Chromatography can also be used to separate colourless substances. These are shown up after chromatography by **spraying** the paper with a **locating** agent. [7]
(1 mark for each word in the correct place)

1.

chromatography / filter paper dipping in solvent [1]
chromatography / filter paper labelled [1]
Solvent labelled [1]
Baseline labelled AND above the level of the solvent [1]

2. a. So that the ink doesn't spread up the paper / graphite / pencil 'lead' doesn't dissolve in solvent [1]
    b. 3 [1]
    c. Ser and Gly [1]
    d. $\dfrac{\text{distance from centre of spot to baseline}}{\text{distance from solvent front to baseline}} = 0.6$ [1]
    e. Label Lys about half way between Cys and Ser/Gly [1]

Unit 2.3

LL The melting and **boiling** points of **pure** substances are sharp. They melt and boil at **exact** temperatures. The melting and boiling points of **impure** substances are not sharp. They melt over a **range** of temperatures. The boiling point of a liquid is **increased** if impurities are present. The melting point of a liquid is **decreased** if impurities are present. [7]
    (1 mark for each word in the correct place)
1. a. Oxygen gas [1] sodium chloride crystals [1]
    b. Small amounts of impurities may react with chemicals [1]
    c. Any values from −15 °C to −1 °C [1]
2. a. Pure sulfur: solidifies at 119 °C
        Pure sulfur: has a sharp boiling point
        Impure sulfur: melts over 4 °C temperature range
        Impure sulfur: turns to a vapour at 450 °C
        (2 marks if all 4 correct; 1 mark if 2 correct) [2]
    b. i. In solder the tin is impure/ the lead is impure [1]
            Impurities lower the melting point [1]
        ii. Less energy is used in melting the solder (than using tin or lead alone) [1]

Unit 2.4

LL A with **3**, B with **4**, C with **5**, D with **1**, E with **2**
    (3 marks if all 5 correct, 2 marks if 3 or 4 correct, 1 mark if 1 or 2 correct) [3]
1. a. i. In order, from top to bottom:
            Filter paper [1]
            Filter funnel [1]
            Flask [1]
        ii. residue on filter paper [1]
            filtrate is liquid in flask [1]
2. a. BGFEADC (2 marks if all correct; 1 mark if 1 pair reversed) [2]
    b. Too much water may dissolve the crystals [1]

Unit 2.5

LL There is a range of **temperatures** in the distillation column, **higher** at the top and **lower** at the bottom. When **vaporised** the more **volatile** alcohols move **further** up the column than the less volatile alcohols. When the alcohol reaches the **condenser** it changes from vapour to **liquid**. The alcohols are collected one by one in the **receiver**, those with the lower **boiling** points condensing before those with higher ones. [10]
    (1 mark for each word in the correct place)
1. a. i. A Distillation flask on left [1]
            B Distillate in beaker on right [1]
            C Slanting tube labelled as condenser [1]
            D Cold water enters the bottom of the condenser [1]
        ii. Arrow under the gauze [1]
    b. i. Salt and water have very different boiling points / salt has a high boiling point and water has a low boiling point [1]
        ii. The vapours would condense together / at the same time [1]
2. a. filtration [1]
    b. simple distillation [1]
    c. fractional distillation [1]

Unit 3.1

LL Atoms are the **smallest** particles of matter which can take part in a **chemical** change. Each atom consists of a **nucleus** made up of protons and **neutrons**. Outside the nucleus are the **electrons**. These are **arranged** in electron **shells** or energy **levels**. [8]
    (1 mark for each word in the correct place)
1. No nucleus in Thomson's model [1]
    Positive charge spread out rather than being in the nucleus [1]
    Electrons not in shells [1]

2. a. i. − / negative [1]
        ii. 0 / no charge [1]
        iii. + / positive [1]
    b. i. The masses using a particular mass as a comparison [1]
        ii. 1/1700 to 1/2000 or 0.0054 or negligible / hardly any mass [1]
3. a. The number of protons / positive charges in the nucleus of an atom [1]
    b. Atomic number [1]
    c. Proton number increases by one across a period [1]

Unit 3.2

LL Isotopes are **atoms** of the same **element** with the same number of **protons** but different numbers of **neutrons**. [4]
    (1 mark each word in the correct place)
1. a. 1 [1]
    b. It has no neutrons [1]
2. [12]

atom / ion of	protons	neutrons	electrons
chlorine	17	18	17
cerium	58	78	58
sodium ion	11	12	10
phosphide ion	15	16	18

3. $^{139}_{57}$La [1]
4. $(185 \times 37.1) + (187 \times 62.9)$ [1]
    18625.8 / 100 ALLOW: answer to 1st step divided by 100 [1]
    186.3 [1]

Unit 3.3

LL The arrangement of the electrons in shells is called the electron **configuration**. An atom of fluorine has nine electrons, **two** in the first shell and **seven** in the second shell. Atoms of elements in the same **group** have the same number of electrons in their **outer** shell. As we move across a **period**, each atom has **one** more **electron** in its outer shell than the element before it. [8]
    (1 mark for each word in the correct place)
1. [8]

element	number of electrons in an atom	electron distribution
nitrogen	7	2,5
oxygen	8	2,6
fluorine	9	2,7
neon	10	2,8
sodium	11	2,8,1
argon	18	2,8,8
calcium	20	2,8,8,2

(1 mark for correct number of electrons in column 1, 1 mark for each correct electron distribution)
2. (1 mark each for correct structure. Electrons should be drawn in shells and paired where possible.) [8]

2,8,3	2,4	2,8,7	2
aluminium	carbon	chlorine	helium
2,8,2	2,8	2,8,5	2,8,8,1
magnesium	neon	phosphorus	potassium

Unit 3.4

LL An element ....is a substance containing only....one type of atom.... [1]
    .... which cannot be broken down further.... by chemical means. [1]
    A compound ....is a substance containing two or more.... types of atom .... [1]
    .... which are chemically .... combined (bonded). [1]

Answers

1. (12 correct = 6, 10 or 11 correct = 5, 8 or 9 correct = 4, 6 or 7 correct = 3, 4 or 5 correct = 2, 2 or 3 correct = 1.) [6]

compound	mixture
The **elements** cannot be **separated** by **physical** means.	The substances in it can be **separated** by **physical** means.
The properties are **different** from those of the **elements** which went to make it.	The properties are the **average** of the substances in it.
The elements are **combined** in a **definite** proportion by mass.	The substances can be **present** in **any** proportion by mass.

2. a. compound [1]
   b. element [1]
   c. element [1]
   d. mixture [1]
   e. compound [1]
   f. mixture [1]

Unit 3.5
LL Malleable; Ductile; Conducts; Lustre; Sonorous; Dense (1 mark each) [6]
1. a. Conducts electricity [1] ductile [1] malleable [1] shiny [1]
   b. i. Poor conductor of electricity / shatters when hit [1]
      ii. Conducts heat / very high melting point [1]
      iii. Conducts electricity / conducts heat / malleable [1]
      iv. Low melting point [1]
2. a. Aluminium [1] has the lowest density [1]
   b. Copper [1] is the best electrical conductor [1]

Unit 4.1
LL A sodium chloride **lattice** is a regular arrangement of **positive** sodium ions and **negative** chloride ions which **alternate** with each other. The ions are held together by **strong** ionic attractive **forces**. This structure is called a **giant** ionic structure. [7]
(1 mark for each word in the correct place)
1. 2,8 ; 2,8,8 ; 2 ; 2,8,18,8 (all correct = 2, 2 or 3 correct = 1) [2]
2.
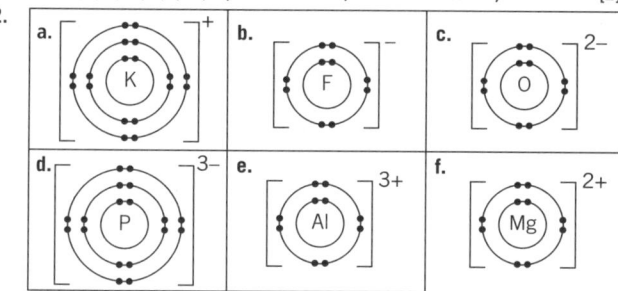
(1 mark each) [6]
3. 8 electrons in outer shell of Mg [1]
   8 electrons in outer shell of S [1]
   2– charge on S [1]

Unit 4.2
LL A with 3, B with 1, C with 4, D with 2 [2]
(2 marks if 4 correct, 1 mark if 2 or 3 correct)
1. CO Cl_2 N_2 O_2 [2]
(2 marks if all correct, 1 mark if 3 correct)
2. a.

hydrogen   bromine

hydrogen bromide   water

(1 mark each) [4]

b. i. 2 around the hydrogen and eight around the others [1]
   (you will not get this mark if you just write eight)
   ii. The electron shells are complete / full [1]
   This is a stable structure / electrons cannot easily be lost or gained [1]

Unit 4.3
LL A covalent bond is formed when **two** atoms combine. It forms because of the **strong** force of **attraction** between the **nucleus** of one atom and the outer **electrons** of the atom next to it. [5]
(1 mark for each word in the correct place)
1.
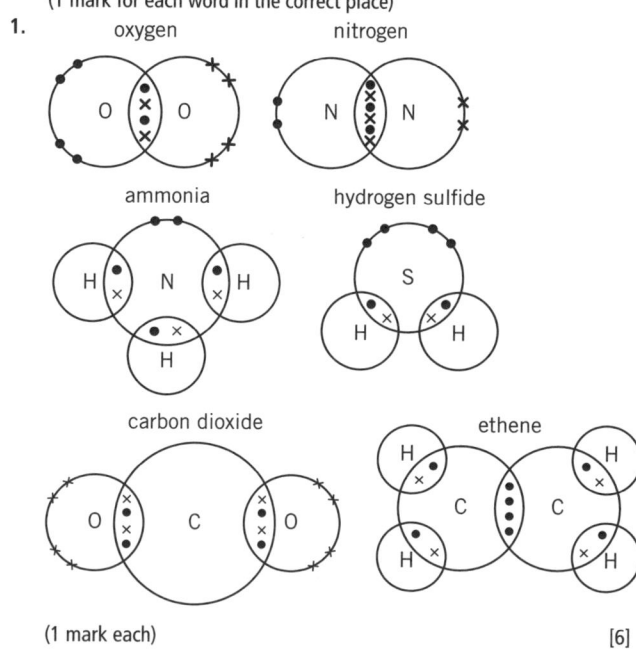
(1 mark each) [6]

Unit 4.4
LL Simple covalent compounds have **low** melting points because the **intermolecular** attractive forces are **weak**. Ionic compounds have **high** melting points because of the **strong** forces of **attraction** between the positive and negative **ions**. [7]
(1 mark for each word in the correct place)
1. Magnesium oxide: ionic
   Carbon tetrachloride: covalent
   Potassium bromide; ionic
   Carbon disulfide: covalent
   Octane: covalent
   (2 marks if all correct and 1 mark if one error) [2]
2. A with **4**; B with **7**; C with **1**; D with **2**; E with **6**; F with **3**; G with **5** [7]
   (All 7 correct = 3 marks, 5 or 6 correct = 2 marks, 3 or 4 correct = 1 mark)

Unit 4.5
LL Layers; Tetrahedral; Giant; Delocalise; Covalent (1 mark each) [5]
1. a. Diamond: **B, C, D** [1]
      Graphite: **A, D, F** [1]
      Silicon dioxide: **B, C, D** [1]
   b. carbon atoms in layers [1] weak forces between layers [1] layers can slide over each other [1]
2. A with **4**; B with **1**; C with **5**; D with **2**; E with **3** [2]
   (all correct = 2 marks, but 3 or 4 correct = 1 mark)
3. Diamond: tetrahedral [1] Graphite: layers of hexagonally arranged / arranged in rings of 6 carbon atoms [1]

Unit 4.6
LL Atoms of metallic elements are generally arranged in closely-packed **layers**. The metal atoms tend to **ionise** and a mobile 'sea' of **electrons** forms between the **positive** metal ions. When a **voltage** is applied to the metal these **delocalised** electrons are able to move. [6]
(1 mark for each word in the correct place)

1. a.

metal ion   free electron

Metal ions shown as + or 2+ [1]
Metal ions labelled [1]
Electrons shown as dots randomly dispersed between the metal ions [1]
Electrons labelled free electrons / delocalised electrons / mobile electrons [1]

b. When force is applied, the force of attraction between metal ions and mobile electrons is weakened [1]
The layers slide over each other [1]
(When force removed) new forces of attraction between metal ions and electrons formed [1]

2. a. D [1]
b. A, B and C [1] Idea that metals are to the left of Group IV / to the left of covalent giant structures (in the Periodic Table) [1]
c. E and F [1] They have low melting points / they are to the right of Period IV [1]

Unit 5.1

LL (1 mark each for any 6 of:) Selenium; Sulfur; Nickel; Carbon; Iodine; Argon; Nitrogen [6]

1. a.

					H 1						
Li 1						B 3	C 4	N 3	O 2	F 1	Ne 0
Na 1	Mg 2					Al 3			S 2	Cl 1	
K 1	Ca 2	transition elements variable / can be different	Zn 2							Br 1	

(1 mark for each column correct (= 9), 1 mark for H, 1 mark for transition elements) [11]
b. Li, Na, K, Mg, Ca, Zn, Al [1]
c. N, O, F, Cl, Br [1]
d. i. H_2S [1]; **ii.** B_2O_3 [1]; **iii.** CS_2 [1]; **iv.** CBr_4 [1]; **v.** Ca_3N_2 [1]; **vi.** Al_2O_3 [1]

2. A magnesium chloride [1] B strontium hydroxide [1]
C iron(II) sulfate [1] D zinc nitrate [1]
E ammonium carbonate [1] F calcium hydrogencarbonate [1]

Unit 5.2

LL A with 1; B with 3; C = 2 (2 marks for all correct; 1 mark for 1 or 2 correct) [2]
1. a. Gain [1]; **b.** Loss of electrons [1]
2. a. A $MgBr_2$ [1] B Na_2O [1] C HCl [1] D $AlCl_3$ [1]
E K_3N [1] F CaS [1] G Al_2S_3 [1] H Fe_2O_3
b. J $Mg(NO_3)_2$ [1] K KOH [1] L $CaHCO_3$ [1] M $(NH_4)_2SO_4$ [1]
N $Ca(OH)_2$ [1] O $Al(NO_3)_3$ P Li_2CO_3 [1]

Unit 5.3

LL A with 3; B with 1; C with 2; D with 4 (2 marks if all correct; 1 mark if 2 or 3 correct) [2]
1. a. $O_2 + 2H_2 \rightarrow 2H_2O$
Correct balance [1] correct use of + and $\rightarrow$ [1]
b. $2C + O_2 \rightarrow 2CO$
Correct balance [1] correct use of + and $\rightarrow$ [1]
2. a. $2K + Br_2 \rightarrow 2KBr$ [1]
b. $4Al + 3O_2 \rightarrow 2Al_2O_3$ [1]
c. $4Na + O_2 \rightarrow 2Na_2O$ [1]
d. $N_2 + 3H_2 \rightarrow 2NH_3$ [1]
e. $2Rb + 2H_2O \rightarrow 2RbOH + H_2$ [1]

Unit 5.4

LL A with 4; B with 3; C with 2; D with 1 (2 marks if all correct, 1 mark if 2 or 3 correct) [2]
1. a. Na^+ and OH^- [1] **b.** Mg^{2+} and Cl^- [1] **c.** Ba^{2+} and NO_3^- [1]
d. Cu^{2+} and SO_4^{2-} [1] **e.** Al^{3+} and O^{2-} [1] **f.** Fe^{2+} and OH^- [1]
2. a. Ions $Cu^{2+} + 2Cl^-$ [1] $(2Na^+) + 2Cl^-$ [1]
Cancel: $Cu^{2+} + \cancel{2Cl^-}$ and $\cancel{(2Na^+)} + \cancel{2Cl^-}$ [1]
Equation: $Cu^{2+}(aq) + 2OH^-(aq) \rightarrow Cu(OH)_2(s)$ [1]
b. Ions: $Ba^{2+} + 2Cl^-$ $Mg^{2+} + SO_4^{2-}$ [1] $\rightarrow Mg^{2+} + 2Cl^-$ [1]
Cancel: $Ba^{2+} + \cancel{2Cl^-}$ $\cancel{Mg^{2+}} + SO_4^{2-}$ [1] $\rightarrow \cancel{Mg^{2+}} + \cancel{2Cl^-}$
Equation $Ba^{2+}(aq) + SO_4^{2-}(aq) \rightarrow BaSO_4(s)$ [1]
c. Ions: $2K^+ + 2I^-$ [1] $\rightarrow 2K^+ + 2Cl^-$ [1]
Equation: $Cl_2(aq) + 2I^-(aq) \rightarrow I_2(aq) + 2Cl^-(aq)$ [1]

Unit 6.1

LL Relative atomic mass (symbol A_r) is the **average** mass of the **isotopes** of an element compared to one-**twelfth** of the mass of an atom of **carbon-12**. [5]
(1 mark for each word in the correct place)

1. a. (2 marks for each correct M_r. If 2 not scored, 1 mark for correct number of atoms) [10]

compound	number of each atom	A_r of atom	M_r calculation
phosphorus trichloride PCl_3	P = 1 Cl = 3	P = 31 Cl = 35.5	$M_r = \dfrac{1 \times 31}{3 \times \mathbf{35.5}}$ 137.5
magnesium hydroxide $Mg(OH)_2$	Mg = 1 O = 2 H = 2	Mg = 24 O = 16 H = 1	$M_r = \dfrac{\begin{array}{l}1 \times 24\\2 \times 16\\2 \times 1\end{array}}{58}$
ethanol C_2H_5OH	C = 2 H = 6 O = 1	C = 12 H = 1 O = 16	$M_r = \dfrac{\begin{array}{l}2 \times 12\\6 \times 1\\1 \times 16\end{array}}{46}$
ammonium sulfate $(NH_4)_2SO_4$	N = 2 H = 8 S = 1 O = 4	N = 14 H = 1 S = 32 O = 16	$M_r = \dfrac{\begin{array}{l}2 \times 14\\8 \times 1\\1 \times 32\\4 \times 16\end{array}}{132}$
glucose $C_6H_{12}O_6$	C = 6 H = 12 O = 6	C = 12 H = 1 O = 16	$M_r = \dfrac{\begin{array}{l}6 \times 12\\12 \times 1\\6 \times 16\end{array}}{180}$

b. i. 342 [1] **ii.** 183 [1] **iii.** 220 [1] **iv.** 130 + 108 = 238

Unit 6.2

LL A with 2, B with 3, C with 4, D with 1 (2 marks if all correct, 1 mark if 2 or 3 correct) [2]

1. (1 mark for each correct answer)

element or compound	formula mass, M_r	mass taken / g	number of moles
O_2	32	4	0.125 [1]
NaCl	58.5 [1]	11.7	0.2 [1]
$CaSO_4$	136 [1]	27.2	0.2 [1]
P_2O_5	142 [1]	56.8 [1]	0.4
CO_2	44 [1]	4.4 [1]	0.1
P_4	124 [1]	86.8	0.7 [1]
CH_4	16 [1]	384 [1]	24.0

2. a. 28 [1] **b.** 20 [1] **c.** 1.204×10^{25} [1]
3. A_r = mass / moles [1] 27 [1] g/mol [1]

Answers

Unit 6.3

LL To find the **number** of moles of a compound, we need to know the mass of compound taken and the **relative** molecular mass of the compound. The relative molecular mass is found by **adding** together the relative **atomic** masses of all the **atoms** (or ions) in the compound. The number of **moles** is found by **dividing** the mass of compound taken by the relative molecular mass. [7]

(1 mark for each word in the correct place)

1. a. $\frac{12}{48}$ [1]   20 g [1]

    b. 4.8 g [1]

    c. $80 \times \frac{168}{48} = 280$ g [1]

2. a. $\frac{2 \times 12}{(2 \times 12) + (6 \times 1)} \times 100 = \mathbf{80}$ **%**

    (2 marks for correct answer, 1 mark if answer wrong but working correct) [2]

    b. $\frac{14}{17} \times 100 = 82.4\%$ (2 marks if correct, 1 mark if answer wrong but working correct) [2]

    c. $\frac{40}{100} \times 100 = 40\%$ (2 marks if correct, 1 mark if answer wrong but working correct) [2]

Unit 6.4

LL **A** with **4**; **B** with **3**; **C** with **1**; **D** with **2** (2 marks if all correct, 1 mark if 2 or 3 correct) [2]

1. $4 \times 34 \rightarrow 4 \times 31 + 6 \times 2$ [1]

    $\mathbf{136}$ g $\rightarrow \mathbf{124}$ g $+ \mathbf{12}$ g [1]

2. a. 5 [1]   b. 1 [1]   c. 2 [1]

    d. mol $I_2O_5 = \frac{24.84}{414} = 0.06$ [1]   So $5 \times 0.06 = 0.30$ mol CO_2 [1]

    e. mol $CO = \frac{21}{28} = 0.75$ [1] So $\frac{0.75}{5} = 0.15$ mol I_2 [1]

3. a. $\frac{168}{56} = 3$ mol [1]

    b. $232 - 168 = 64$ g O [1] $\frac{64}{16} = 4$ mol O [1]

    c. 3Fe:4O [1] Fe_3O_4 [1]

Unit 6.5

LL At room **temperature** and pressure, one **mole** of any gas occupies 24 **dm³** (24000 cm³). Because there is the same **number** of moles, there is also the same number of gas **molecules** in a given volume. So, 50 cm³ of chlorine and 50 cm³ of **oxygen** under the same **conditions** contain the same number of molecules. [7]

(1 mark for each word in the correct place)

1. (1 mark for each correct answer) [10]

gas	M_r of gas	mass of gas / g	moles of gas / mol	volume of gas / dm³
ammonia	17	8.5	0.5 [1]	12 [1]
oxygen	32	64 [1]	2 [1]	48
carbon dioxide	44	3.08	0.07 [1]	1.68 [1]
hydrogen chloride	36.5 [1]	292	8	192 [1]
ethane	30	3.75 [1]	0.125 [1]	3

2. a. $\frac{48}{24000} = 2 \times 10^{-3}$ mol H_2 [1]

    $\frac{25}{24000} = 1.042 \times 10^{-3}$ mol O_2 [1]

    b. 2×10^{-3} mol H_2 reacts with 1×10^{-3} mol O_2 [1]

    1.042×10^{-3} mol O_2 is greater than 1×10^{-3} mol O_2 so oxygen is in excess and hydrogen limiting [1]

    c. 2×10^{-3} mol H_2 forms 2×10^{-3} mol H_2O [1]

    $2 \times 10^{-3} \times 18 = 0.036$ g [1]

Unit 6.6

LL **percentage** purity $= \dfrac{\text{mass of } \underline{\text{pure}} \text{ product}}{\text{mass of } \underline{\text{impure}} \text{ product}} \times 100$ [4]

(1 mark for each word in the correct place)

1. a. 100 g / mol [1]

    b. 3.840 dm³ [1]

    c. $\frac{3.840}{24} = 0.16$ mol [1]

    d. 0.16 mol [1]

    e. $0.16 \times 100 = 16$ g [1]

    f. $\frac{16}{18} (\times 100) = 89$ % [1]

2. a. 122 g / mol [1]

    b. $\frac{24.4}{122} = 0.2$ mol [1]

    c. 0.2 mol [1]

    d. 136 g / mol [1]

    e. $0.2 \times 136 = 27.2$ g [1]

    f. $\frac{25.84}{27.2} (\times 100) = 95$ % [1]

Unit 6.7

LL The molecular formula of a **compound** shows the number of **each** type of atom in one **molecule**. The empirical formula shows the **simplest** ratio of **atoms** which **combine**. [6]

(1 mark for each word in the correct place)

1. moles of Pb = **0.1 mol**   moles of Cl = **0.4 mol** [1]

    Divide by lowest number of moles   Pb $\frac{\mathbf{0.1}}{\mathbf{0.1}}$ Cl $\frac{\mathbf{0.4}}{\mathbf{0.1}}$ [1]

    Result of division = **1**   = **4**

    Simplest ratio **1Pb:4Cl**. [1] So empirical formula is $\mathbf{PbCl_4}$ [1]

2. a. HO [1]   b. Sb_2O_3 [1]   c. C_2H_5 [1]

3. A empirical formula mass 110 [1]   molecular formula P_4O_6 [1]

    B empirical formula mass 67.5 [1]   molecular formula S_2Cl_2 [1]

    C empirical formula mass 30 [1]   molecular formula $C_2H_4O_2$ [1]

Unit 6.8

LL concentration in **moles** per dm³ $= \dfrac{\text{amount of } \underline{\text{solute}} \text{ in moles}}{\text{volume in } \underline{\text{dm}^3}}$ [5]

(1 mark for each word in the correct place)

1.

solute	M_r of solute	mass of solute / g	volume of solution cm³ or dm³	concentration of solution mol/dm³
sodium hydroxide	40	8	250 cm³	0.48 [1]
silver nitrate	170	17 [1]	200 cm³	0.5
copper(II) sulfate	160	40	2.0 dm³ [1]	0.125

2. a. moles of acid $= \underline{0.10} \times \frac{12.2}{1000} = \mathbf{1.22 \times 10^{-3}}$ mol H_2SO_4 [1]

    b. i. 2 (1)   ii. $1.22 \times 10^{-3} \times 2 = 2.44 \times 10^{-3}$ mol NaOH [1]

    c. $\frac{2.44 \times 10^{-3}}{0.025} = 0.098$ mol / dm³ [2]

    (2 marks for correct answer, 1 mark for 0.025 if correct answer not obtained)

Unit 7.1

LL Across: 1. DECOMPOSE; 5. ELECTRODE; 6. ION; 7. BATTERY;

Down: 2. CATHODE; 3. EDONA (ANODE in reverse); 4. CELL

(1 mark for each correct word) [7]

1. Aluminium is a good conductor of electricity [1]

    Aluminium has low density so cable less likely to break under its own weight [1]

    Iron core gives the cable extra strength [1]

2. A: battery/ cell(s) / power supply [1]

    B: anode [1]

    C: cathode [1]

    D: electrolyte [1]

Answers

3. a. Electrolysis: breakdown of ionic substance when molten or in
      solution [1]
      by passage of electricity [1]
   b. Electrolyte: liquid which conducts electricity [1]
   c. Anode: Positive electrode [1]
4. a. Conducts electricity   [1]   inert / unreactive [1]
   b. graphite / carbon   [1]   platinum [1]

Unit 7.2

LL When dilute sulfuric acid is electrolysed using **platinum** electrodes we
    observe **bubbles** at each electrode. Hydrogen is formed at the **cathode**
    and oxygen is produced at the **anode**. We can think of this electrolysis
    being the electrolysis of **water**. The hydrogen is formed from hydrogen
    ions and the oxygen from **hydroxide** ions. [7]
    (1 mark for each correct word)

1. (1 mark each 'cell' correct) [12]

electrolyte	cathode (−) product	anode (+) product	observations at the anode
Molten zinc bromide	zinc	bromine	red-brown fumes / solution
Molten magnesium chloride	magnesium	chlorine	yellow-green / green fumes
Molten calcium oxide	calcium	oxygen	bubbles (colourless)
Molten lead(II) iodide	lead	iodine	brown solution / purple vapour

2. a. $2(NaCl)$ [1] $2 (Na)$ [1] $Cl_2(g)$ [1]
   b. $2(Mg)$ [1] O_2 [1]
   c. PbI_2 [1] $[I]$ [1]
3.

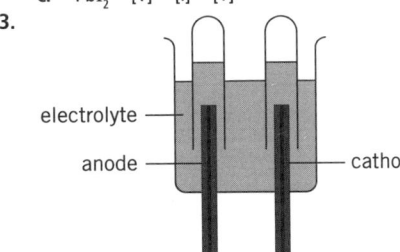

   a. test tube over each electrode [1] and some liquid in the test tube [1]
   b. + electrode labelled 'anode' and − electrode labelled 'cathode' [1]
   c. surface of electrolyte shown above the level of the electrodes [1]
      electrolyte labelled 'electrolyte' or 'hydrochloric acid' [1]

Unit 7.3

LL If a metal .... is more reactive than hydrogen, ... the metal ions stay in
    solution and .... hydrogen arising from .... hydrogen ions in water, ...
    bubbles off. [2]
    (2 marks if all correct, 1 mark if one pair of phrases in the incorrect
    place)

1. (1 mark for each 'cell' correct) [15]

electrolyte	cathode (−) product	anode (+) product	observations at the anode
concentrated KCl(aq)	hydrogen	chlorine	bubbles of gas, green when collected
dilute H_2SO_4(aq)	hydrogen	oxygen	colourless bubbles
dilute NaCl(aq)	hydrogen	chlorine	colourless bubbles
concentrated HCl(aq)	hydrogen	chlorine	bubbles of gas, green when collected
dilute KI(aq)	hydrogen	iodine	brown solution / purple vapour

2. a. B chlorine   [1]   C hydrogen   [1]   b. A [1]
   c. Chlorine is more reactive than oxygen. Chlorine discharged more
      readily [1]
      than oxygen from hydroxide ions in water [1]

Unit 7.4

LL During electrolysis, positive ions move towards the **cathode** where they
    gain **electrons**. This is a **reduction** reaction. Negative ions **move** to the
    anode where they **lose** electrons. This is an **oxidation** reaction. In the
    external circuit the electrons travel in the wires from the **negative** pole
    of the battery to the cathode. [8]
    (1 mark for each word in the correct place)

1. a.   molten zinc bromide   b.   dilute sulfuric acid

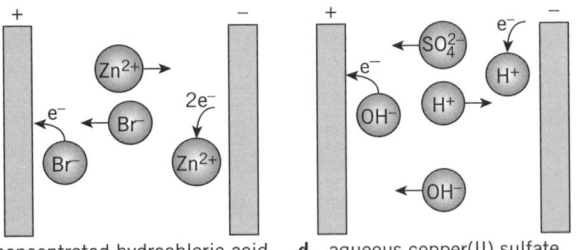

   c. concentrated hydrochloric acid   d. aqueous copper(II) sulfate

   a. Zinc ions moving to cathode and bromide ions to anode [1]
   b. Bromide ions donating electrons to anode and zinc ions taking
      electrons from cathode [1]

2. a. $Zn^{2+} + \underline{2e^-} \rightarrow Zn$ [1]
   b. $\underline{2}Cl^- \rightarrow \underline{Cl_2} + \underline{2}e^-$ (1 mark for 2 and Cl_2, 1 mark for balance
      with electrons) [2]
   c. $\underline{2}H^+ + \underline{2}e^- \rightarrow \underline{H_2}$ (1 mark for 2 and H_2, 1 mark for balance with
      electrons) [2]
   d. $Al^{3+} + \underline{3}e^- \rightarrow \underline{Al}$ [1]
   e. $\underline{4}OH^- \rightarrow \underline{O_2} + \underline{2}H_2O + \underline{4}e^-$ (1 mark for correct species, 1 mark
      for balance) [2]

Unit 7.5

LL The electrolysis cell has an **impure** strip of copper as the **anode** and
    a pure strip of **copper** as the cathode. The **electrolyte** is a solution of
    copper(II) sulfate. At the anode, copper **atoms** lose electrons and go
    into **solution** as copper(II) **ions**. At the cathode, copper(II) ions **gain**
    electrons and are deposited on the **cathode** as copper atoms. [9]
    (1 mark for each word in the correct place)

1. a. Anode: $Cu \rightarrow \underline{Cu^{2+}} + \underline{2}e^-$ (1 mark for correct species, 1 mark
         for balance) [2]
         Cathode: $\underline{Cu^{2+}} + \underline{2}e^- \rightarrow Cu$ (1 mark for correct species, 1 mark for
         balance) [2]
   b. Cu^{2+} ions removed from cathode and Cu^{2+} formed at the anode [1]
      At same rate [1]

2.

mass of the electrodes	anode: no change	[1]	anode: decreases	[1]
	cathode: increases slightly	[1]	cathode: increase	[1]
appearance	anode: none / bubbles given off	[1]	anode: gets thinner	[1]
	cathode: goes pink / brown	[1]	cathode: gets thicker with lighter colour pink deposit	[1]
electrolyte	gets a lighter blue/ fades	[1]	remains the same depth of colour	[1]

Unit 7.6

LL The object to be **electroplated** is connected to the **negative** pole
    of the **power** supply. The object becomes the **cathode**. A strip of the
    plating **metal** is connected to the positive **pole** of the power supply.
    The **plating** metal is the anode. The **electrolyte** is a solution containing
    ions of the plating metal. [9]
    (1 mark for each word in the correct place)

1. a. A = rod, B = jug, C = liquid / solution in which the anode and
        cathode dip   [2]
        (3 correct = 2 marks, 1 or 2 correct = 1 mark)
    b. gains mass (slightly) [1] becomes silvery [1]
    c. anode: $Ag \rightarrow \underline{Ag^+} + \underline{e^-}$ [1]
        cathode $\underline{Ag^+} + \underline{e^-} \rightarrow \underline{Ag}$ [1]
2. a. Tin forms a layer over the iron / tin covers the iron   [1]
        Prevents water or oxygen from getting to the surface of the iron [1]
    b. Makes them look attractive   [1]

Unit 7.7
LL Bauxite; Oxygen; Alumina; Cryolite; Carbon (1 mark each)   [5]
1. E = liquid in the cell into which rods are dipping   [1]
    C = layer on the inside next to liquid   [1]
    A = rods dipping into the liquid   [1]
    M = layer at the bottom of the cell   [1]
2. Aluminium oxide has a very high melting point   [1]
    Cryolite / calcium fluoride dissolves the aluminium oxide   [1]
    Melting point of the electrolyte is lowered   [1]
3. a. i. $Al^{3+} + \underline{3e^-} \rightarrow \underline{Al}$ (1 mark for correct formulae, 1 mark for
            balance)   [2]
        ii. $\underline{2}O^{2-} \rightarrow O_2 + \underline{4e^-}$ (1 mark for correct formulae, 1 mark for
            balance)   [2]
    b. $2Al_2O_3 \rightarrow 4Al + 3O_2$ (1 mark for correct formulae, 1 mark for
        balance)   [2]

Unit 7.8
LL Ceramics are useful .... insulators, not only .... because they resist
    the flow ..... of electricity, but also because .... they have very high
    ... melting points.   [2]
    (2 marks if all correct, 1 mark if one pair of phrases in the incorrect
    place)
1. a. i. (indicator) bulb/ lamp labelled [1] cells / battery / power source
            labelled   [1]
        ii. Arrows in direction carbon rod to + of battery or − of battery to
            bulb/ carbon rod   [1]
2. a. good   [1]
    b. poor   [1]
    c. poor   [1]
    d. good   [1]
    e. good   [1]
    f. poor   [1]
3. A with 3; B with 1; C with 4; D with 2 (2 marks if 4 correct, 1 mark if 2
    or 3 correct)   [2]

Unit 8.1
LL A with 4; B with 3; C with 1; D with 2 (2 marks if all correct, 1 mark if
    2 or 3 correct)   [2]
1. The following should be underlined: Separating iron from sulfur [1],
    melting zinc [1] Distilling plant oils [1]
2. a. endothermic   [1]
    b. exothermic   [1]
    c. endothermic   [1]
    d. exothermic   [1]
3. a. surroundings   [1]
    b. increases   [1]
    c. solution   [1]
    d. enthalpy   [1]
    e. thermal   [1]
    f. faster   [1]

Unit 8.2
LL A with 4; B with 1; C with 2; D with 3 (2 marks if all correct, 1 mark if
    2 or 3 correct)   [2]
1. a. **Energy** on the vertical axis of both **L** and **M**   [1]
        Reactants on the lines on the left of both **L** and **M** [1]
        Products on the lines on the right of both **L** and **M** [1]
        Downward arrow between the two lines in **L**   [1]
        Upward arrow between the two lines in **M**   [1]

b. The energy of the reactants is greater than the energy of the
    products   [1]
    So energy is released   [1]
c.

    Axes correctly labelled   [1]
    Reactant and products correct with products above reactants
    in terms of energy   [1]
    Arrow for enthalpy change correct   [1]
    Arrow for activation energy correct   [1]
    Arrows correctly labelled   [1]

Unit 8.3
LL In an endothermic reaction, the energy **absorbed** in bond breaking is
    greater than the energy **released** when new **bonds** are formed. In an
    exothermic reaction, the energy given **out** when **new** bonds are formed
    is greater than the energy taken in when the bonds in the **reactants**
    are broken.   [7]
    (1 mark for each word in the correct place)
1.

bonds broken (endothermic +) / kJ		bonds formed (exothermic −) / kJ	
$4 \times (C–H) = 4 \times 413 = \underline{1652}$	[1]	$2 \times (C=O) = 2 \times 805 = 1610$	[1]
$\underline{2} \times (O=O) = \underline{2 \times 498} = \underline{996}$	[1]	$4 \times (O–H) = 4 \times 464 = 1856$	[1]
Total	$+ 2648$	Total =	$−3466$

Overall energy change = (+2648) + (−3466) = −818 kJ   [1]

2.

bonds broken (endothermic +) / kJ		bonds formed (exothermic −) / kJ	
$(H–H) = 435.9$		$2 \times (H–Cl) = 2 \times 432 = 864$	[1]
$(Cl–Cl) = 243.4$			
Total $+ 679.3$	[1]	Total =	$−864$

Overall energy change = (+679.3) + (−864) (1) = −138.7 kJ   [1]

Unit 8.4
LL A with 3; B with 1; C with 2 (2 marks if all correct, 1 mark if 1 or
    2 correct)   [2]
1. a. D   [1]
    b. B   [1]
    c. i. B [1] it has the lowest density / it weighs least   [1]
        ii. Hazard of the fuel e.g. how combustible it is / how poisonous
            it is / its state / whether it is solid/ liquid or gas   [1]
2. a. $\underline{2}H_2 + O_2 \rightarrow \underline{2H_2O}$ (1 mark for correct symbols, 1 mark for balance)
    b. $\underline{2}C_2H_6 + \underline{7O_2} \rightarrow \underline{4CO_2} + \underline{6H_2O}$ (1 mark for correct symbols, 1 mark
        for balance)   [2]
    c. $C_7H_{16} + \underline{11O_2} \rightarrow \underline{7CO_2} + \underline{8H_2O}$ (1 mark for correct symbols, 1 mark
        for balance)   [2]
    d. $\underline{2}H_2S + \underline{3O_2} \rightarrow \underline{2H_2O} + \underline{2}SO_2$ (1 mark for correct symbols, 1 mark
        for balance)   [2]
3. air / oxygen [1] incomplete [1] monoxide [1]

Unit 8.5

LL A fuel cell consists of two **porous** electrodes coated with **platinum**. The electrolyte is either an acid or an **alkali**. Hydrogen and **oxygen** are bubbled through the porous electrodes where the **reactions** take place. When connected to an **external** circuit, **electrons** flow from the **negative** electrode to the positive electrode. [8]
(1 mark for each word in the correct place)

1. a. voltmeter [1]
    b. D [1]
    c. A [1]
2   a. $2H_2 + 4OH^- \rightarrow 4H_2O + 4e^-$ (1 mark for correct symbols, 1 mark for balance) [2]
    b. $O_2 + 2H_2O + 4e^- \rightarrow 4OH^-$ (1 mark for correct symbols, 1 mark for balance) [2]
3. Advantages: 1 mark each for any two of: no pollutants formed or only water formed / lighter in weight / more efficient or fewer moving parts / produces more energy per g of fuel burnt [2]
    Disadvantages: 1 mark each for any two of: hydrogen more flammable / hydrogen is mainly made from fossil fuels (at present) / difficulties with storing gas [2]

Unit 9.1

LL To find the rate of reaction we can either measure how **quickly** the reactants are **used** up or how quickly the **products** are formed. To calculate the **rate** of reaction we need to find out how some measurement changes with **time**. For example, the **volume** of gas given off per **second**, or how the mass of the reaction mixture **decreases** with time. [8]
(1 mark for each word in the correct place)

1. B, C, D, A [1]
2. a. Copper reducing in size [1]
        Solution getting darker [1]
        Gas given off [1]
    b. (1 mark each for any three of:) decrease in mass of copper per minute; increase in volume of gas per minute; increase in depth of colour of solution / increase in copper compound per minute; decrease in concentration of acid per minute [3]
3. Electrical conductivity (in this reaction) due to ions [1] Ions on the left but none on right [1]

Unit 9.2

LL A with 3; B with 4; C with 1; D with 2 (2 marks if all correct, 1 mark if 2 or 3 correct) [2]

1. a. Test tube not graduated [1] replace by measuring cylinder [1]
        Gas from reaction mixture can escape into air [1] provide stopper with tube going through it [1]
        Delivery tube dipping into acid so most of gas not getting to measuring cylinder [1] make tube near the top of the flask in the stopper [1]
        Test tube empty at start so gas volume cannot be measured [1]
        Fill inverted measuring cylinder to top with water [1]
    b. Stopclock / timer [1]
2. Increase in gas volume because hydrogen has low density ALLOW: hydrogen very light [1] Difficult to measure very small changes in mass with an ordinary balance (even one reading to two decimal places) [1]

Unit 9.3

LL At the start of the reaction, the .... volume of hydrogen given off .... per second is high. As the reaction proceeds, the volume .... of hydrogen given off per second decreases .... which is shown by the gradient .... of the graph decreasing.
(2 marks if all correct, 1 mark if one sentence correct or one pair of errors) [2]

1. a. 66 s [1]
    b. 46 cm³ [1]
    c. 34 cm³ [1]

d. Points all plotted correctly [1] Best curve through the points [1]

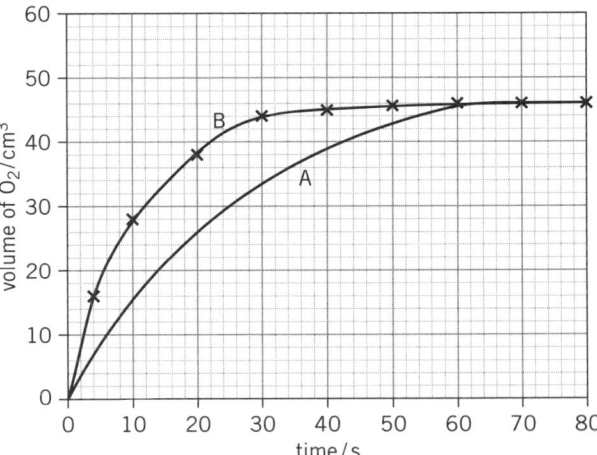

Unit 9.4

LL A catalyst is a substance which **increases** the **rate** of a reaction. The catalyst remains **unchanged** at the **end** of the reaction. A catalyst works by providing a route for the **reaction** which has a **lower** activation energy. [6]
(1 mark for each word in the correct place)

1. a. 24 cm³ [1]
    b. 48 cm³ [1]
    c. The smaller ones [1] More ions / particles exposed for reaction [1]
2. a. Three curves similar to that in Unit 9.3 all starting at 0–0 and levelling off at the same volume [1]
        Steepest curve labelled S and shallowest curve labelled L [1]
    b.

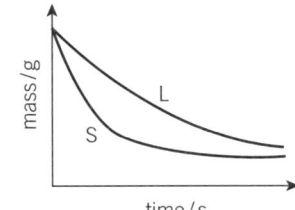

Horizontal axis labelled time with units of s or min [1]
Vertical axis labelled mass with units of g [1]
2 curves levelling off [1]
Steepest curve labelled S / shallowest curve labelled L [1]

Unit 9.5

LL In order to react, particles must **collide** with each other. The collisions must have enough **energy** to break **bonds** to allow a reaction to happen. Increasing the concentration of a reactant **increases** the **frequency** of collisions and so increases the **rate** of reaction. [6]
(1 mark for each word in the correct place)

1. a.

More acid particles drawn [1]
Same number Mg particles remaining [1]
Particles randomly spread out including water particles [1]
    b. Fewer acid particles drawn [1]
        Fewer magnesium particles drawn [1]
        Particles drawn randomly including water particles (and $MgCl_2$ particles) [1]
2. (When pressure is increased) molecules are closer [1] So they collide more frequently [1]

Answers

Unit 9.6

LL A with **3**; B with **4**; C with **1**; D with **2** (2 marks if all correct, 1 mark if 2 or 3 correct) [2]

1. **a.**

Correct axes labelled with units [1] Full use of grid [1] 7 points plotted correctly [2] (but 5 or 6 plotted correctly = [1]) Curved line of best fit [1]

b. Increases with increase in temperature [1] Comment that increase gets increasingly greater / greater with each 10 °C rise [1]

Unit 10.1

LL A with **3**; B with **4**; C with **2**; D with **1** (2 marks if all correct, 1 mark if 2 or 3 correct) [2]

1. **a.** Reversible reaction / equilibrium reaction [1]
b. Warm / heat gently [1]
c. Heating hydrated cobalt chloride to get anhydrous cobalt chloride is endothermic [1]
So the reverse reaction must be exothermic [1]

2. **a.** No matter can enter or exit [1] closed container / closed tube [1]
b. Concentrations of reactants and products are constant [1] Rate of forward reaction equals rate of backward reaction [1]
c. The extent to which the reaction is in favour of reactants or products [1]

Unit 10.2

LL If a reaction is exothermic in the forward direction, it will be **endothermic** in the reverse direction. For an exothermic reaction, when temperature increases, the equilibrium **shifts** in the direction of the **reverse** reaction. It **favours** the endothermic change where **heat** is taken in. [5]
(1 mark for each word in the correct place)

1. **a.** left [1]
b. left [1] more [1] left [1] OR fewer [1] right [1] NOTE: third mark dependent on second being correct [1]
c. right

2. **a.** to the right [1]
b. to the left [1]
c. to the right [1]
d. no effect [1]

3. **a.** White precipitate disappears [1]
b. Increase in amount of white precipitate [1]

Unit 10.3

LL A with **3**; B with **1**; C with **2**; D with **4** (2 marks if all correct, 1 mark if 2 or 3 correct) [2]

1. **a.** Arrow from hydrogen to water labelled oxidation [1]
Arrow from oxygen to water labelled reduction [1]
b. Arrow from lead oxide to lead labelled reduction [1]
Arrow from hydrogen to water labelled oxidation [1]
c. Arrow from iron oxide to iron labelled reduction [1]
Arrow from carbon to carbon monoxide labelled oxidation [1]
d. Arrow from carbon to carbon monoxide labelled oxidation [1]
Arrow from methane to hydrogen labelled reduction [1]

e. Arrow from zinc oxide to zinc labelled reduction [1]
Arrow from carbon to carbon monoxide labelled oxidation [1]
f. Arrow from iron to iron oxide labelled oxidation [1]
Arrow from water to hydrogen labelled reduction [1]

Unit 10.4

LL A with **2**; B with **3**; C with **4**; D with **1** (2 marks if all correct, 1 mark if 2 or 3 correct) [2]

1. **a.** Magnesium oxidised and oxygen reduced [1]
b. Hydrogen oxidised and lead reduced [1]
c. Iodide ion oxidised and chlorine reduced [1]
d. Iodide ion oxidised and oxygen reduced [1]

2. **a.** $Ca \rightarrow Ca^{2+} + 2e^-$ [1] oxidation [1]
b. $Cl_2 + 2e^- \rightarrow 2Cl^-$ [1] reduction [1]
c. $Al^{3+} + 3e^- \rightarrow Al$ [1] reduction [1]
d. $Fe^{2+} \rightarrow Fe^{3+} + e^-$ [1] oxidation [1]
e. $O_2 + 4e^- \rightarrow 2O^{2-}$ [1] reduction [1]
f. $Pb^{4+} + 2e^- \rightarrow Pb^{2+}$ [1] reduction [1]
g. $2Br^- \rightarrow Br_2 + 2e^-$ [1] oxidation [1]

3. **a.** −1 [1] **b.** −3 [1] **c.** +6 [1]
d. +3 [1] **e.** +5 [1] **f.** −2 [1]

Unit 10.5

LL A with **3**; B with **4**; C with **2**; D with **1** (2 marks if all correct, 1 mark if 2 or 3 correct) [2]

1. **a.** Oxidising agent is O_2, reducing agent is Mg [1]
b. Oxidising agent is PbO, reducing agent is H_2 [1]
c. Oxidising agent is Cl_2, reducing agent is I^- [1]
d. Oxidising agent is H_2O_2, reducing agent is I^- [1]

2. **a.** colourless to purple [1]
b. potassium manganate(VII) [1] Its oxidation number decreases / its oxidation number goes from +7 to +2 [1]

3. **a.** Add aqueous potassium iodide to aqueous iron(III) ions [1]
yellow solution ALLOW colourless solution [1] turns brown [1]
b. Iodine [1] I_2 [1]

Unit 11.1

LL Across: 1. SULFUR; 2. ALKALINE; 5. LITMUS; 7. INDICATOR;
Down: 1. SOAP; 2. ACID; 3. NEUTRAL; 4. VERS (uniVERSal); 6. ONE
(1 mark for each word) [9]

1. A with **3**; B with **5**; C with **4**; D with **1**; E with **2** [3]
(3 marks if all correct, 2 marks if 3 or 4 correct, 1 mark is 1 or 2 correct)

2. **a.** (1 mark each)
i. NH_3 ii. CH_3COOH iii. H_2CO_3 iv. H_2SO_4
v. $Ca(OH)_2$ vi. H_3PO_4 vii. HNO_3 viii. NaOH
b. Universal indicator [1]

3. Can 'eat away' at the surface of another substance [1]

Unit 11.2

LL Acids react with many metals to form a salt and **hydrogen**. A salt is a substance formed when particular hydrogen **atoms** in an acid are **replaced** by a metal. Acids also react with some metal **hydroxides** to produce a **salt** and water and with carbonates to produce a salt, **water** and carbon **dioxide**. [7]
(1 mark for each word in the correct place)

1. **a.** pH 13 [1]   **b.** pH 7 [1]   **c.** pH 3 [1]

2. **a.** zinc oxide + hydrochloric acid → zinc chloride + water [1]
b. iron + sulfuric acid → iron sulfate + hydrogen [1]
c. sulfuric acid + lead carbonate → lead sulfate + carbon dioxide + water [1]
d. hydrochloric acid + tin oxide → tin chloride + water [1]

3. **a.** $Zn + H_2SO_4 \rightarrow ZnSO_4 + H_2$ [1]
b. $MgO + 2HNO_3 \rightarrow Mg(NO_3)_2 + H_2O$ (1 for correct formula, 1 for balance) [2]
c. $CuCO_3 + 2HCl \rightarrow CuCl_2 + CO_2 + H_2O$ (1 for correct formula, 1 for balance) [2]
d. $Na_2CO_3 + 2HCl \rightarrow 2NaCl + CO_2 + H_2O$ (1 for correct formula, 1 for balance) [2]
e. $Ca + 2HCl \rightarrow CaCl_2 + H_2$ (1 for correct formula, 1 for balance) [2]
f. $2KOH + H_2SO_4 \rightarrow K_2SO_4 + 2H_2O$ (1 for correct formula, 1 for balance) [2]

4. Hydrogen ion [1] H^+ [1]

Answers

Unit 11.3

LL A base is a substance which reacts with an **acid** to form a salt. Bases which are **soluble** in water are called **alkalis**. Many metal **oxides** are basic because they react with acids to form a **salt** and water. Ammonia reacts with **sulfuric** acid to form ammonium **sulfate**; so ammonia is also a base. [7]
(1 mark for each word in the correct place)

1. a. sodium hydroxide + nitric acid → sodium nitrate + water [1]
   b. calcium oxide + hydrochloric acid → calcium chloride + water [1]
   c. barium hydroxide + nitric acid → barium nitrate + water [1]
2. a. $MgO + 2HNO_3 → Mg(NO_3)_2 + H_2O$ (1 for correct formula, 1 for balance) [2]
   b. $2NaOH + H_2SO_4 → Na_2SO_4 + 2H_2O$ (1 for correct formula, 1 for balance) [2]
   c. $Ca(OH)_2 + 2HNO_3 → Ca(NO_3)_2 + 2H_2O$ (1 for correct formula, 1 for balance) [2]
   d. $2NH_3 + H_2SO_4 → (NH_4)_2SO_4$ (1 for correct formula, 1 for balance) [2]
3. $Ca(OH)_2 + 2NH_4Cl → CaCl_2 + 2H_2O + 2NH_3$ (1 for correct formula, 1 for balance) [2]
4. Soil may be acidic (because of addition of fertilisers) [1] Calcium oxide / carbonate neutralise the acidity [1]
5. OH^- ions are formed [1]
6. a. hydroxide ion [1] OH^- [1]
   b. $H^+ + OH^-$ [1] $→ H_2O$ [1]

Unit 11.4

LL Aqueous solutions of acids contain **hydrogen** ions. In strong acids the acid **molecules** ionise (**dissociate**) completely to form hydrogen **ions** and anions. When weak acids dissolve in **water** they become **partially** dissociated. [6]
(1 mark for each word in the correct place)

1. Acid: proton donor [1] Base: proton acceptor [1]
2. a. pH: ethanoic = pH 2.9 [1] hydrochloric = pH 1.0 [1] sulfuric = pH 0.7 [1]
   b. Rate of reaction: ethanoic slow [1] hydrochloric fast [1] methanoic slow [1] sulfuric fast [1]
3. a. $H^+(aq) + \cancel{NO_3^-(aq)} + \cancel{Na^+(aq)} + OH^-(aq) → \cancel{NO_3^-(aq)} + \cancel{Na^+(aq)} + H_2O(l)$ [1]
   b. $H^+(aq) + OH^-(aq) → H_2O(l)$ [1]
   c. Hydrogen ions have reacted completely with hydroxide ions to form water [1]

Unit 11.5

LL A with 4; B with 3; C with 2; D with 1 (2 marks if all correct, 1 mark if 2 or 3 correct) [2]
1. a. red / pink [1] blue [1] b. pink / red [1] yellow [1]
   c. colourless [1] blue [1]
2. a. 11.2 [1] b. 21 cm³ [1]
   c. pH decreases slowly at first [1] then rapidly [1] then slowly (when excess acid added) [1]

Unit 11.6

LL A with 2; B with 3; C with 1 (2 marks if all correct, 1 mark if 1 or 2 correct) [2]
1. a. $MgO + 2HCl → MgCl_2 + H_2O$ (1 for correct formula, 1 for balance) [2]
   b. $SO_2 + 2NaOH → Na_2SO_3 + H_2O$ (1 for correct formula, 1 for balance) [2]
   c. $CuO + H_2SO_4 → CuSO_4 + H_2O$ [1]
   d. $CO_2 + 2NaOH → Na_2CO_3 + H_2O$ (1 for correct formula, 1 for balance) [2]
   e. $ZnO + 2HNO_3 → Zn(NO_3)_2 + H_2O$ (1 for correct formula, 1 for balance) [2]
   f. $CaO + H_2SO_4 → CaSO_4 + H_2O$ [1]
2. a. $SO_2 + H_2O → H_2SO_3$ [1]
   b. $CO_2 + H_2O → H_2CO_3$ [1]
   c. $CaO + H_2O → Ca(OH)_2$ [1]
   d. $P_4O_6 + 6H_2O → 4H_3PO_3$ (1 for 6H_2O, 1 for 4H_3PO_3) [2]
   e. $Na_2O + H_2O → 2NaOH$ (1 for NaOH, 1 for balance) [2]
3. $ZnO + 2KOH → K_2ZnO_2 + H_2O$ (1 for K_2ZnO_2, 1 for H_2O) [2]

Unit 12.1

LL Filtrate; Insoluble; Filter; Crystal; Evaporate; Heat; Oxide (1 mark each) [5]
1. a. Filter off the excess zinc [1]
   b. Filter off the crystals [1]
   Wash them in the filter paper with a minimum amount of solvent / alcohol [1]
   Dry the crystals between sheets of filter paper / dry in a **drying** oven / dry in the air [1]
2. DBEAFC [2]
   (1 mark if one pair reversed)
3. a. $CaO + H_2SO_4 → CaSO_4 + H_2O$ (1 mark for H_2SO_4, 1 mark for rest correct) [2]
   b. $Zn + 2HCl → ZnCl_2 + H_2$ (1 mark for HCl, 1 mark for rest correct) [2]

Unit 12.2

LL Across: 1. TITRE; 3. SOLUTE; 6. INDICATOR; 7. ACID; 8. NINE; 9. END; Down: 2. TITRATION; 4. BURETTE; 5. PIPETTE
(1 mark for each correct word) [9]
1. Record the volume of acid added when the indicator just changed colour [1]
   Repeat the titration without the indicator, using the value of acid recorded [1]
   Evaporate to the point of crystallisation and leave to form crystals [1]
   Filter off crystals and wash with solvent [1]
   Dry crystals with filter paper / dry in **drying** oven [1]
2. a. $NaOH + HNO_3 → NaNO_3 + H_2O$ (1 mark for HNO_3, 1 mark for rest correct) [2]
   b. $2NH_3 + H_2SO_4 → (NH_4)_2SO_4$ (1 mark for H_2SO_4, 1 mark for rest correct) [2]

Unit 12.3

LL Salts such as **nitrates**, sodium salts and **ammonium** salts are soluble in water. Many **carbonates / hydroxides** and **carbonates / hydroxides** are insoluble except those from Group I. An insoluble substance formed when two **solutions** of soluble **compounds** are mixed is called a **precipitate**. [7]
(1 mark for each word in the correct place)
1. a. soluble [1] b. soluble [1] c. soluble [1]
   d. insoluble [1] e. insoluble [1] f. insoluble [1]
2. BEADC (1 mark if one pair in the incorrect order) [2]
3. a. i. $Ag^+(aq) + \cancel{NO_3^-(aq)} + \cancel{K^+(aq)} + Br^-(aq) → AgBr(s) + \cancel{NO_3^-(aq)} + \cancel{K^+(aq)}$ [1]
   ii. $Ag^+(aq) + Br^-(aq) → AgBr(s)$ [1]
   b. $Ba^{2+}(aq) + SO_4^{2-}(aq) → BaSO_4(s)$ [2]
   (1 mark for corrections, 1 mark for state symbols)

Unit 12.4

LL When a gas is denser than **air**, you collect it by **upward** displacement of air. If a gas is **less** dense than air, you **collect** it by **downward** **displacement** of air. [6]
(1 mark for each word in the correct place)
1. a. i. C [1] ii. B [1] iii. A [1]
   b. i. A [1] ii. D [1]
2. A with 2; B with 4; C with 1; D with 3 (2 marks if all correct, 1 mark if 2 or 3 correct) [2]

Unit 12.5

LL A **sample** is put on the end of a platinum **wire** and placed at the **edge** of a non-**luminous** Bunsen **flame**. If **lithium** is present, the flame is **coloured** red. If **potassium** is present, the flame has a lilac colour. [8]
(1 mark for each word in the correct place)
1. a. pale green [1]
   b. orange-red [1]
   c. blue-green [1]
   d. yellow [1]
2. $Al^{3+}(aq)$ with sodium hydroxide: white precipitate [1]
   which dissolves in excess [1]
   with ammonia: white precipitate [1] insoluble in excess [1]
   $Cr^{3+}(aq)$ with sodium hydroxide: green precipitate [1] which dissolves in excess [1]

with ammonia: grey-green precipitate [1] insoluble in excess [1]

Cu^{2+}(aq) with sodium hydroxide: light blue precipitate [1]
insoluble in excess [1]

with ammonia: light blue precipitate [1] dissolves in excess to
form a dark blue solution [1]

Fe^{3+}(aq) with sodium hydroxide: red-brown precipitate [1] which is
insoluble in excess [1]

with ammonia: red-brown precipitate [1] insoluble in excess [1]

Unit 12.6

LL A few **drops** of nitric acid are added to the **solution** thought to be a
halide.

Aqueous silver **nitrate** is then added. If a **chloride** is present, a
white **precipitate** is seen. If a bromide is present, a **cream**-coloured
precipitate is seen. [6]
(1 mark for each word in the correct place)

1. A with 4; B with 1; C with 3; D with 2 (2 marks if all correct, 1 mark if
2 or 3 correct) [2]

2. a. $AgNO_3$(aq) + NaCl(aq) → AgCl(s) + $NaNO_3$(aq)
(1 mark for correct formulae, 1 mark for state symbols) [2]
   b. Ag^+(aq) + Cl^-(aq) → AgCl(s) [1]

3. a. $BaCl_2$(aq) + Na_2SO_4(aq) → $BaSO_4$(aq) + 2NaCl(aq)
(1 mark for correct formulae, 1 mark for balance, 1 mark for state
symbols) [3]
   b. Ba^{2+}(aq) + SO_4^{2-}(aq) → $BaSO_4$(s) [1]

Unit 13.1

LL Across: 4. Sr; 5. CHLORINE; 6. BORON; 8. NICKEL;
Down: 1. MAGNESIUM; 2. Ar; 3. SODIUM; 4. SULFUR; 5. CARBON; 7. AL
(1 mark for each correct word) [10]

1. a. 2nd line: 2,8,2; 2,8,3; 2,8,4; 2,8,5; 2,8,6; 2,8,7 [1]
4th line: Al_2O_3 [1] SiO_2 [1] P_2O_3 [1]
   b. i. They increase to a maximum at Si then decrease [2]
(They increase then decrease = 1 mark)
      ii. metallic [1]
      iii. It is a giant structure / has a giant lattice [1]
All the bonds are strong / its takes a lot of energy to break all
the bonds [1]
      iv. They are simple molecules [1]
Weak forces between (the molecules) / it doesn't take much
energy to break the weak intermolecular forces [1]

Unit 13.2

LL The **alkali** metals have low **melting** points and **densities** compared
with most other metals. They react with water to form a metal
hydroxide and **hydrogen**. [5]
(1 mark for each word in the correct place)

1. a. Melting point of potassium ALLOW between 50 and 75°C
(actual = 63 °C) [1]
Metallic radius of rubidium ALLOW between 0.21 and 0.24 nm
(actual = 0.235) [1]
Observations
Sodium: (1 mark each for any three of:)
Any value between 1.7 and 2.0 g / cm^3 (actual = 1.88)
Moves rapidly over the surface / Fizzes rapidly / Melts and goes
into a ball / Does not burst into flame [3]
Rubidium: (1 mark each for any three of:)
Whizzes over the surface or moves faster than potassium / Fizzes
extremely rapidly or fizzes more than potassium / Bursts into flame
immediately / May explode [3]
   b. Any value between 1.7 and 2.0 g / cm^3 (actual = 1.88) [1]

Unit 13.3

LL When aqueous chlorine is added to a **colourless** solution of potassium
bromide, the solution turns **orange** because **bromine** has been
displaced. This is because a **more** reactive **halogen** displaces a **less**
reactive halogen from an aqueous solution of its **halide**. [7]
(1 mark for each correct word)

1. a. Melting point increases down the Group [1]
   b. fluorine: gas [1] chlorine: liquid [1] bromine: solid [1] iodine:
solid [1]

c. Arrow going downwards from light to dark [1]
d. Arrow going downwards from smaller to larger [1]

2. Orange solution turns red-brown [1] because iodine is formed [1]
Bromine is more reactive than iodine [1]
A more reactive halogen displaces a less reactive halogen from
its halide [1]

Unit 13.4

LL The Group VIII gases (**noble** gases) are unreactive because their
electron configurations make them **stable**. It is difficult for their **atoms**
to form ionic bonds by **gaining** or losing electrons or covalent bonds
by **sharing** electrons. Helium is particularly unreactive because there
cannot be **more** than two electrons in the first **shell**. The other gases
have **eight** electrons in their outer shell which is a stable electronic
configuration. [9]
(1 mark for each correct word)

1. They have a complete / stable outer shell of electrons [1]
So they cannot share, gain or lose electrons to form diatomic
molecules [1]

2. a. They increase to a maximum at Si then decrease [2]
(They increase then decease = 1 mark)
   b. (1 mark each for any 4 of:)
(Na) Mg and Al are metals so have high melting points because of
attraction between + ions and sea of electrons [1]
Melting points increase from Na to Al because more outer shell
electrons donated to the sea of electrons [1]
So greater attraction between electrons and metal ions [1]
Silicon has highest melting since it is a giant structure / has
a giant lattice [1]
All the bonds are strong / its takes a lot of energy to break all the
bonds [1]
P, S and Cl have relatively low melting points because they are
simple molecules [1]
Weak forces between (the molecules) / it doesn't take much
energy to break the weak intermolecular forces [1]
   c. Si, P, S, Cl poor conductors because they have no delocalised
electrons [1]
Na, Mg, Al good conductors because the delocalised electrons can
move between the metal ions [1]
   d. Melting points increase from Na to Al because more outer shell
electrons donated to the sea of electrons [1] So greater number of
delocalised electrons able to move [1]

Unit 13.5

LL Across: 2. COB; 4. COLOURED; 6. ON; 8. ETIC; 9. NI; 10. CATALYST;
Down: 1. IRON; 2. CHROME; 3. SOFT; 4. COPPER; 5. DENSITY; 7. POINT
(1 mark for each correct word) [12]

1. a. A, D, E, G, H (3 marks if all correct, 2 marks if 4 correct, 1 mark
if 3 correct) [3]
   b. (1 mark each for any two of:)
very hard / very strong [1] form complex ions [1] form ions with
different charges [1]

2. a. Ag^+ [1]   b. Cu^{2+} [1]   c. Cr^{3+} [1]   d. Fe^{3+} [1]

3. Fe^{3+}(aq) + 3OH^-(aq) → $Fe(OH)_3$(s)
(1 mark for correct formulae, 1 mark for balance, 1 mark for state
symbols) [3]

Unit 14.1

LL The products formed by metals which react with **cold** water are a metal
hydroxide and **hydrogen**. The hydroxides are **alkaline** and so turn red
litmus **blue**. The products formed by metals which only react with steam
are a metal **oxide** and hydrogen. Copper does not react with water
because it is **lower** in the reactivity series than hydrogen and cannot
take the **electrons** away from the hydrogen in the water. [7]
(1 mark for each correct word)

1. a. 2Na(s) + 2H_2O(l) → 2NaOH(aq) + H_2(g)
(1 mark for H_2, 1 mark for balance, 1 mark for state symbols) [3]
   b. 3Fe(s) + 4H_2O(g) → Fe_3O_4(s) + 4H_2(g)
(1 mark for formulae, 1 mark for balance, 1 mark for
state symbols) [3]

Answers

2. a. calcium, magnesium, zinc [1]
    b. barium: gives off bubbles very rapidly with <u>cold</u> water [1]
        barium disappears very quickly / immediately [1]
        lead reacts slowly (when white hot) with steam / no reaction
        even when heated [1]
3. lead, iron, magnesium, lithium (1 mark if one pair reversed / all reversed) [2]

Unit 14.2

LL Although aluminium is high in the <u>reactivity</u> series, samples of the
    metal exposed to the air do not appear to react with water or dilute
    <u>acids</u>. This is because <u>freshly</u>-made aluminium reacts with <u>oxygen</u> in
    the air to form a thin <u>layer</u> of aluminium <u>oxide</u> on its surface, which
    is relatively <u>unreactive</u>. The oxide layer <u>sticks</u> to the metal surface
    strongly so is not easily removed. [8]
    (1 mark for each correct word)
1. a. reducing agent Fe and oxidising agent CuO [1]
    b. reducing agent Mg and oxidising agent Fe_2O_3 [1]
2. manganese oxide + aluminium → manganese + aluminium oxide [1]
3. a. $SnO_2 + 2C → Sn + 2CO$
        (1 mark for correct formulae, 1 mark for balance) [2]
    b. $2NiO + CO + H_2 → 2Ni + CO_2 + H_2O$
        (1 mark for correct formulae, 1 mark for balance) [2]
    c. $2PbO + C → 2Pb + CO_2$
        (1 mark for correct formulae, 1 mark for balance) [2]
4. a. silver < copper < tin < chromium < manganese
        (2 marks if all correct, 1 mark if 1 pair reversed) [2]
    b. $Cr_2O_3 + 3C → 2Cr + 3CO$
        (1 mark for correct formulae, 1 mark for balance) [2]

Unit 14.3

LL Metals above carbon in the reactivity series are not usually <u>extracted</u>
    from their oxides by heating with <u>carbon</u>. This is because the metal
    <u>bonds</u> to the oxygen too strongly. Carbon is not <u>reactive</u> enough
    to remove the oxygen from the metal oxide unless a very high
    <u>temperature</u> is used. Iron which is below carbon in the reactivity series
    is extracted from <u>hematite</u> ore. Aluminium, which is above carbon in
    the reactivity series is extracted by <u>electrolysis</u> of aluminium oxide
    obtained from <u>bauxite</u> ore. [8]
    (1 mark for each correct word)
1. a. sodium; magnesium; aluminium; iron; copper; gold
        (all correct = 3 marks; 4 or 5 correct = 2 marks; 2 or
        3 correct = 1 mark) [3]
    b. potassium, sodium; calcium; magnesium; aluminium [1]
        They are above hydrogen in the reactivity series [1]
2. a. $Fe_2O_3 + 3CO → 2Fe + 3CO_2$ (If 2 not scored, 1 mark for [2]
        $2Fe + 3CO_2$)
    b. $ZnO + CO → Zn + CO_2$ [1]
    c. $2CuO + C → 2Cu + CO_2$ [1]
3. carbon monoxide [1]

Unit 14.4

LL Calcium is <u>higher</u> in the reactivity series than copper. So calcium atoms
    are better at <u>losing</u> electrons than copper. When calcium is added to
    <u>aqueous</u> copper(II) sulfate, calcium <u>displaces</u> the copper and copper
    metal is formed. Calcium <u>atoms</u> are converted to calcium ions which go
    into <u>solution</u>. [6]
    (1 mark for each correct word)
1. a. B At start: solution <u>blue</u> [1]
        B After 20 min: metal <u>brown/ pink</u> [1] solution <u>colourless /
        lighter blue</u> [1]
        C After 20 min: metal <u>silvery-grey</u> [1]
        D At start: metal grey [1] solution <u>blue</u> [1]
        D After 20 min: metal <u>brown/ pink</u> [1] solution <u>colourless /
        lighter blue</u> [1]
    b. silver < copper < iron < zinc [1]
    c. No reaction because copper is lower in the reactivity series
        than zinc / no reaction because zinc is higher than copper in the
        reactivity series [1]

Unit 15.1

LL Hematite; Reduction; Slag; Oxide; Iron; Blast; Carbon; Coke [8]
    (1 mark each)

1.
    B [1]
    E ← → E [1]
    A → ← A [1]
    D [1]   C [1]

2. Burning coke in air to form carbon dioxide [1]
    Reduction of carbon dioxide with coke to form carbon monoxide [1]
3. Limestone undergoes <u>thermal</u> decomposition to form calcium <u>oxide</u>.
    This reacts with <u>silicon</u> dioxide (sand) which is an <u>impurity</u> in the ore.
    The calcium <u>silicate</u> (<u>slag</u>) formed runs down the furnace and <u>floats</u>
    on top of the <u>molten</u> iron. [8]
    (1 mark for each correct word)
4. $Fe_2O_3 + 3CO → 2Fe + 3CO_2$
    (1 mark for correct formulae, 1 mark for balance) [2]

Unit 15.2

LL Blocks of zinc can be placed on the hull of a ship to stop it <u>rusting</u>. Zinc
    is <u>more</u> reactive than <u>iron</u> so it loses <u>electrons</u> and forms <u>ions</u> more
    easily than iron. The zinc ions go into <u>solution</u> and so the zinc <u>corrodes</u>
    instead of the iron. This is called <u>sacrificial</u> protection. [8]
    (1 mark for each correct word)
1. a. Corrosion increases greatly at low pH [1]
        Corrosion decreases at very alkaline pH [1]
        Not much difference in corrosion between pH 3 and 9 [1]
    b. hydrated [1] iron(III) oxide [1]
2. a. There is very little water / water evaporates very quickly [1]
    b. Idea of a layer protecting the surface of the iron [1]
        So prevents air / oxygen and water from reaching the iron [1]
    c. Layer of zinc [1] stops air / oxygen and water reaching the iron [1]
        Zinc more reactive than iron [1] so corrodes in preference to iron [1]

Unit 15.3

LL A with 4; B with 1; C with 3; D with 2 (all correct = 2 marks; 2 or 3
    correct = 1 mark) [2]
1. B [1]
2. Alloys are <u>mixtures</u> of metals or mixtures of metals with non-metals.
    Alloys are often <u>harder</u> and stronger than pure metals. When a metal
    is <u>alloyed</u> with another metal, the <u>difference</u> in the size of the metal
    atoms makes the arrangement of the <u>layers</u> in the lattice less <u>regular</u>.
    This <u>prevents</u> the layers from sliding over each other as easily when a
    <u>force</u> is applied. [8]
    (1 mark for each correct word)
3. Aluminium alloy: (1 mark each for any 2 of:) low density or lightweight /
    doesn't corrode / strong (for weight) [2]
    Stainless steel: (1 mark each for any two of:) rusts or corrodes less /
    stronger / harder [2]
4. Idea of not a pure metal [1] Impurities lower the melting point [1]

Unit 15.4

LL A with 4; B with 3; C with 1; D with 2 (all correct = 2 marks; 2 or
    3 correct = 1 mark) [2]
1. Aluminium food containers: (1 mark each for any 2 of:) non-toxic (oxide)
    / resistant to corrosion / low density or lightweight [2]
    Aluminium power cables: low density [1] conducts electricity [1]
    Copper for electrical wiring: good conductor of electricity [1] ductile [1]
    Stainless steel: Use (1 mark each for any 2 of:) e.g. chemical plant /
    cutlery / surgical instruments [3]
        Property: resistant to corrosion / very hard [1]
    Tungsten steel: (1 mark each for any 2 of:) e.g. resistant to
    wear / high melting point / hard [3]
    Brass: attractive or shiny [1] hard or strong [1]

Unit 16.1

LL Nitrogen; Iron; Catalyst; Haber; Methane; Hydrogen; Pressure [7]
(1 mark each)

1. Methane and steam [1]
2. $N_2(g) + 3H_2(g) \rightleftharpoons 2NH_3(g)$ (1 mark for correct formulae, 1 mark for balance) [2]
3. a. Increasing pressure increases % yield [1]
   b. The % yield decreases with increasing temperature [1]
   c. 52% [1]
   d. Advantage: rate of reaction faster [1]
      Disadvantage: % yield lower [1]

Unit 16.2

LL For healthy growth crop plants need three major elements, nitrogen, **phosphorus** and potassium. Plants take up these elements in the form of nitrates, **phosphates** and potassium **salts**. The **nitrates** are needed to make **proteins** for growth. Farmers add **fertilisers** to the soil to add back the **nutrients** which plants have absorbed for growth. [7]
(1 mark for each correct word)

1. a. NH_3 ammonia [1] HNO_3 nitric acid [1] NH_4NO_3 ammonium nitrate [1]
      H_2SO_4 sulfuric acid [1] H_3PO_4 phosphoric acid [1] KCl potassium chloride [1]
   b. ammonia + nitric acid $\rightarrow$ ammonium nitrate [1]
   c. i. ammonia [1] sulfuric acid [1]
      ii. potassium hydroxide / potassium carbonate [1] hydrochloric acid [1]
          NOT: potassium
      iii. Sodium hydroxide / sodium carbonate [1] phosphoric acid [1]
          NOT: sodium
2. a. To neutralise acid in the soil [1]
   b. $(NH_4)SO_4 + Ca(OH)_2 \rightarrow CaSO_4 + 2NH_3 + 2H_2O$
      (1 mark for correct formulae, 1 mark for balance) [2]

Unit 16.3

LL Vanadium; Catalyst; Trioxide; Oleum; Exothermic; Oxygen; Contact [7]
(1 mark each)

1. a. catalyst / to speed up the rate of reaction [1]
   b. i. As temperature increases from 300 to about 450 °C there is not much difference in yield / the yield gets a little less [1]
      At temperatures higher than about 450 °C the yield decreases markedly [1]
      ii. 92–93 % [1]
      iii. For an exothermic reaction the yield decreases as temperature increases [1]
   c. Increases the yield of product / shifts the equilibrium to the right [1]
      There are fewer moles / smaller volume of gas on the right in the equation [1]

Unit 16.4

LL When fossil fuels containing sulfur are **burnt**, sulfur **dioxide** gas is formed. Sulfur dioxide reacts with water **vapour** in the atmosphere to form **sulfurous** acid. Some of the sulfur dioxide is **oxidised** in the atmosphere to form sulfur trioxide. This reacts with water vapour to produce **sulfuric** acid. Acid rain has a **pH** of about 4.5 and so can harm wildlife in **lakes** and damage the **leaves** of trees. [9]

1. a. lightning / high temperature furnaces / car exhausts / bacterial action in soil [1]
   b. nitrous acid / nitric acid [1]
2. During the morning there are more vehicles on the road [1]
   Nitrogen dioxide from vehicle exhausts goes into the atmosphere [1]
   Nitrogen dioxide concentration builds up during the day since there is still traffic on the roads [1]
   After about 12 midnight very few vehicles about so the nitrogen oxides have time to disperse into the upper atmosphere [1]
3. Chemical erosion / eats the surface away [1] NOT dissolves
   Acids react with carbonates [1] to form carbon dioxide, calcium chloride and water [1]

Unit 17.1

LL A with 1; B with 3; C with 4; D with 2 (all correct = 2 marks; 2 or 3 correct = 1 mark) [2]

1. a. 0.04%   b. 78%   c. about 1%   d. oxygen 21%
      (1 mark for each) [4]

2. Carbon monoxide is formed when **carbon**-containing compounds **burn** in a **limited** supply of air. Sulfur dioxide is formed when **fossil** fuels containing **sulfur** burn in air. [5]
   (1 mark for each correct word)
3. a. global warming / stated effect of global warming e.g. melting ice caps [1]
   b. toxic / poisonous / stops respiration [1]
4. a. carbon dioxide AND methane [1]
   b. Water (vapour) [1]

Unit 17.2

LL Across: 2. SYNTHESIS; 3. RESPIRATION; 6. FUEL; 7. SIX; 8. GLUCOSE; Down: 1. DECAY; 2. SHELLS; 4. OCEANS; 5. COAL
(1 mark for each correct word) [9]

1. a. Respiration is (slightly more than) balanced by photosynthesis [1]
   b. Cutting down trees / removing vegetation [1] Burning more fossil fuels [1]
2. carbon dioxide + water $\rightarrow$ glucose + oxygen (1 mark for correct reactants 1 mark for correct products) [2]
   $6CO_2 + 6H_2O \rightarrow C_6H_{12}O_6 + 6O_2$
   (1 mark for correct species, 1 mark for correct balance) [2]

Unit 17.3

LL A with 4; B with 3; C with 2; D with 1 (all correct = 2 marks; 2 or 3 correct = 1 mark) [2]

1. a. Gas which absorbs energy heat / infrared radiation [1] in the atmosphere [1]
   b. The general trend is the same of increasing concentration of CO_2 and increasing temperature of the atmosphere [1] Reference especially to the years 1950 to 2000 [1]
2. (1 mark each for any 2 of:) e.g. rise in sea level /desertification / more extreme weather / melting glaciers / warming of sea causing death of corals etc. [2]
3. Methane is a greenhouse gas which is formed by the **bacterial** decomposition of **vegetation** and as a waste product of **digestion** in animals. It is present in the **atmosphere** at a lower concentration than carbon dioxide but it **absorbs** much more thermal energy per mole. [5]
   (1 mark for each correct word)

Unit 17.4

LL A with 3; B with 4; C with 1; D with 2 (all correct = 2 marks; 2 or 3 correct = 1 mark) [2]

1. a. To turn harmful carbon monoxide [1] and nitrogen oxides [1] into nitrogen and carbon dioxide which are not harmful (to health) [1]
   b. i. $2NO_2 \rightarrow N_2 + 2O_2$ (1 mark for correct products, 1 mark for balance) [2]
      ii. $2NO + 2CO \rightarrow N_2 + 2CO_2$ (1 mark for correct products, 1 mark for balance) [2]
2. Flue gas desulfurisation is the process of removing **sulfur** dioxide from the gases formed during the **combustion** of fossil **fuels** in furnaces. The **waste** gases are passed through moist calcium **carbonate** or calcium oxide. These compounds **neutralise** the acidic sulfur dioxide. Solid calcium **sulfite** is formed. [7]
   (1 mark for each correct word)
3. a. Climate change is the change in the climate / weather over a number of years due to global warming [1]
      Causing more extreme weather / example of more extreme weather e.g. ore storms / more heatwaves [1]
      Important to reduce its effect otherwise get e.g. desertification *causing* loss of land for agriculture / loss of homes / habitats *caused* by sea level rise [1]
   b. i. Increases photosynthesis [1] so more carbon dioxide absorbed [1]
      ii. Reduces the use of fossil fuels [1] so less carbon dioxide gets into the atmosphere [1]

Unit 17.5

LL In a water treatment **plant**, large objects such as plant **branches** are first trapped by metal screens. Other solid particles are then left to **settle** to the bottom of the tank. The water is then passed through a **filter** made of sand or gravel. This removes small **insoluble** particles. Carbon is added to remove bad **smells**. Chlorine is added to the filtered water to kill **bacteria** which may be **harmful** to health. [8]
(1 mark for each correct word)

Answers

1. Oxygen: respiration of aquatic organisms [1]
   Mineral salts: for health / correct functioning of enzymes in aquatic organisms [1]
2. a. Ca^{2+} [1]
   b. NO_3^- [1] and SiO_3^{2-} [1] and K^+ [1] present in much higher concentration in river water
   c. 2.4 mg [1]

Unit 17.6
LL A with 2; B with 3; C with 1; D with 6; E with 4; F with 5 (all correct = 3 marks; 4 or 5 correct = 2 marks; 2 or 3 correct = 1 mark) [3]
1. Plastics can get into the ocean from ships, coastal towns and by transport in **rivers** from inland. Plastic fishing nets can **trap** or kill **fish** and other sea creatures. Very small particles of plastics called **microplastics** have been found in drinking water. These particles are so **small** that they can get into our bloodstream and then to organs such as the **liver** and kidney where they may cause harm. [6]
   (1 mark for each correct word)
2. A with 4; B with 3; C with 1; D with 2 (all correct = 2 marks; 2 or 3 correct = 1 mark) [2]

Unit 18.1
LL A with 3; B with 4; C with 2; D with 1 (2 marks if all correct, 1 mark if 2 or 3 correct) [2]
1. a. A family of similar compounds with similar chemical properties [1] due to the same functional group [1]
   b. propene: alkenes [1]
      butanol: alcohols [1]
      hexane: alkanes [1]
      propanoic acid: carboxylic acids [1]
2. ethane: C_2H_6 [1]
   ethanol: C_2H_6O [1]
   ethanoic acid: $C_2H_4O_2$ [1]
   ethene: C_2H_4 [1]
3. a. [structure of methanol]   b. [structure of methanoic acid]
   (1 mark for each structure correct) [2]
4. a. $C_nH_{2n+1}COOH$ / $C_nH_{2n+1}CO_2H$ [1]
   b. C_nH_{2n+2} [1]
   c. C_nH_{2n} [1]
   d. $C_nH_{2n+1}OH$ [1]

Unit 18.2
LL A with 4; B with 3; C with 1; D with 2 (all correct = 2 marks; 2 or 3 correct = 1 mark) [2]
1. a. [structure]   b. [structure]
   c. [structure]   d. [structure]
   e. [structure]   f. [structure]
   (1 mark each) [6]
2. a. C_3H_7- [1] b. CH_3- [1] c. $C_6H_{13}-$ [1] d. C_2H_5- [1]

Unit 18.3
LL A structural formula is the simplest **unambiguous** (clearly defined) description of how the **atoms** are arranged in a molecule. It shows each **carbon** atom in a molecule together with the **number** of the other atoms

attached to each carbon atom. Compounds with the same **molecular** formula but different **structural** formulae are called **isomers**. [7]
(1 mark for each correct word)
1. a. $CH_3CH_2CH_2CH_3$ [1]
   b. CH_3CH_2OH [1]
   c. CH_3CO_2H [1]
   d. CH_3CHCH_2 [1]
   e. $CH_3CH(OH)CH_3$ [1]
   f. $CH_3CH_2CH_3$ [1]
2. a. [two structures]
   (1 mark for each isomer) [2]
3. [structures]
   (1 mark each for any two isomers) [2]

Unit 18.4
LL Across: 4. VOLATILE; 5. WATER; 6. CU; 7. DIESEL; 9. SOLID;
    Down: 1. COAL; 2. METHANE; 3. PETROL; 4. VISCOUS; 8. ERY (refinERY)
    (1 mark for each correct word) [10]
1. a. C [1] b. B [1]
2. a. Downward arrow [1]
   b. i. Easily vaporised / liquid has a low boiling point [1]
      ii. Upward arrow [1]
   c. Downward arrow [1]
   d. Upward arrow [1]

Unit 18.5
LL There is a range of **temperatures** in the distillation column, hot at the **bottom** and cooler at the **top**. Hydrocarbons with **lower** boiling points move **further** up the column and **condense** when the temperature in the column falls just below the **boiling** point of the hydrocarbons. Hydrocarbons with **higher** boiling points condense lower down the column. [8]
(1 mark for each correct word)
1. Heat the flask with flame of constant height [1]
   Collect first fraction in test tube over a particular temperature range [1]
   Replace test tube and continue to heat until another fraction is collected over a particular temperature range [1]
   Repeat until no more fractions can be collected [1]

Answers

2. **A** with **4**; **B** with **3**; **C** with **2**; **D** with **1** (all correct = 2 marks; 2 or 3 correct = 1 mark) [2]

Unit 19.1
LL Alkanes are generally **unreactive** except for the reaction with chlorine in the presence of light (a **photochemical** reaction) and **combustion**. When **excess** alkane is mixed with chlorine in a sealed tube and exposed to **sunlight**, the **green** colour of the chlorine disappears. A chlorine atom replaces a **hydrogen** atom in the alkane. This type of reaction is called a **substitution** reaction. If excess **chlorine** is present, more than one hydrogen atom is replaced by chlorine. [9]
(1 mark for each correct word)

1. a. hydrocarbons [1] b. single [1] covalent [1]
2. Boiling points increase as relative molecular mass increases [1]
3. a. $C_5H_{12} + 8O_2 \rightarrow 5CO_2 + 6H_2O$
   (1 mark for balancing carbon dioxide and water, 1 mark for balancing oxygen) [2]
   b. $2C_4H_{10} + 9O_2 \rightarrow 8CO + 10H_2O$
   (1 mark for balancing butane, carbon monoxide and water, 1 mark for balancing oxygen) [2]
   c. $CH_4 + Cl_2 \rightarrow CH_3Cl + HCl$ [1]
   d. $C_2H_6 + 2Cl_2 \rightarrow C_2H_4Cl_4 + 2HCl$ [1]
   (1 mark for HCl as product, 1 mark for balance) [2]
4. photochemical [1] substitution [1]
5.

(1 mark for each) [2]

Unit 19.2
LL Across: 1. CRACKING; 3. ETHENE; 5. HYDROGEN; 7. THERMAL; Down: 1. CATALYST; 2. NINE; 4. HEAT; 6. OXIDE
(1 mark for each correct word) [8]

1. a. residues [1]
   b. (1 mark for name and 1 mark for structure of any of:)
   methane, CH_4; ethane, C_2H_6; propane, C_3H_8, butane, C_4H_{10} [2]
   c. i. gasoline / naphtha and diesel [1]
   ii. kerosene and fuel oil and residue [1]
2. a. $C_{10}H_{22} \rightarrow C_4H_{10} + C_6H_{12}$ [1]
   b. $C_{14}H_{30} \rightarrow C_3H_8 + C_4H_8 + C_7H_{14}$ [1]

Unit 19.3
LL We can tell the difference between an unsaturated and a **saturated** compound by adding **aqueous** bromine to a sample of the compound. Aqueous bromine is **orange** in colour. If the aqueous bromine is **decolourised** the compound is **unsaturated**. If the aqueous bromine **remains** orange, the compound is saturated. [6]
(1 mark for each correct word)

1. a. alkenes: propene [1] butene [1]
   molecular formulae: C_2H_4 [1] C_4H_8 [1] C_5H_{10} [1]
   boiling points: propene −80 to −20 °C (actual −48 °C) [1]
   hexene 40 to 90 °C (actual 63 °C) [1]
   b. pentene and hexene [1]
2. C=C ringed [1]
3. **A**
   B $H_2O(g)$ (NOT $H_2O(l)$) [1]
   (1 for formula, 1 for state symbol) [2]

Unit 19.4
LL Ethanol is a colourless **liquid** at r.t.p. It burns in excess **air** to produce carbon **dioxide** and **water**. Ethanol can be oxidised to **ethanoic** acid by oxygen from the air in the presence of **bacteria**. Enzymes from the bacteria **catalyse** the reaction. In the laboratory, ethanol can be converted to ethanoic acid by oxidation with **acidified** potassium **manganate (VII)**. [9]
(1 mark for each correct word)

1. a. arrow under flask pointing upwards [1]
   b. oxidising agent [1]
   c. purple [1] to colourless [1]
   d. To condense the vapours / to stop loss of vapour [1] of ethanol / ethanoic acid [1]
2. a. $C_2H_5OH + 3O_2 \rightarrow 2CO_2 + 3H_2O$
   (1 mark for correct formulae, 1 mark for balance) [2]
   b. $C_2H_5OH + 2[O] \rightarrow CH_3COOH + H_2O$
   (2 marks if all formulae correct, 1 mark if two of C_2H_5OH, CH_3COOH, H_2O correct) [3]
3. solvent [1] fuel [1] ALLOW: making esters

Unit 19.5
LL Anaerobic; Ethanol; Enzymes; Yeast; Glucose; Distil (1 mark each) [6]
1. a. fermentation: reagents glucose [1] temperature ALLOW between 10 and 40 °C [1] pressure atmospheric / 1 atm [1] catalyst enzymes / yeast [1]
   hydration: reagents ethene and steam [1] temperature 500–600 °C [1] pressure 60–70 atmospheres [1] catalyst phosphoric acid [1]
   b. (1 mark each for any two of:) takes a long time / ethanol is dilute / need to distil off the ethanol / batch process is inefficient / lot of waste [2]
   c. (1 mark each for any two of:) uses renewable resources / relatively cheap / does not require very high temperature and pressure [2]
   d. (1 mark for any of:) made from non-renewable source / petroleum / needs expensive equipment / needs more expensive catalyst [1]
   e. (1 mark each for any two of:) reaction is fast / reaction can be run continuously / gives pure ethanol or atom economy is (nearly) 100% [2]

Unit 19.6
LL **A** with **3**; **B** with **4**; **C** with **1**; **D** with **2** (all correct = 2 marks; 2 or 3 correct = 1 mark) [2]
1. a. The reaction is an equilibrium reaction [1] Both un-ionised acid molecules and (ethanoate) ions are present [1]
   b. It is accepting a proton [1] from ethanoic acid [1]
2. a. $2CH_3COOH + 2Na \rightarrow 2CH_3COO^-Na^+ + H_2$
   (1 mark for formula of each product ALLOW CH_3COONa, 1 mark for balance) [2]
   b. $2CH_3COOH + Mg \rightarrow 2(CH_3COO^-)_2Mg^{2+} + H_2$
   (1 mark for formula of each product ALLOW $(CH_3COO)_2Mg$, 1 mark for balance) [2]
   c. $CH_3COOH + NaOH \rightarrow CH_3COO^-Na^+ + H_2O$
   (1 mark for each correct product) [2]
   d. $CH_3COOH + CH_3OH \rightarrow CH_3COOCH_3 + H_2O$
   (1 mark for each correct reactant and 1 mark for H_2O) [2]
3. a.
   [1]
   b.
   [1]
4. a. butyl methanoate [1]
   b. propyl propanoate [1]

Unit 20.1
LL A polymer is a substance which has very large **molecules** formed when lots of small molecules called **monomers** join together. This process is called **polymerisation**. When poly(ethene) is formed, one of the C=C **bonds** of **ethene** is broken and the monomers **join** together in a chain.
(1 mark for each correct word) [6]

1. 8 carbon atoms in a chain with two hydrogen attached to each [1]
   Continuation bonds shown on the end carbon atoms [1]
2. They cannot be broken down / be decomposed [1] by living organisms / bacteria / fungi [1]
3. a. Saves raw material / saves energy / can be made into new objects / could be cracked to different chemicals [1]
   b. Toxic gases produced [1]

173

4. PET polymers are **hydrolysed** (broken down using **aqueous** acids or alkalis) to their **monomers**. The monomers are then **repolymerised**. [4]
(1 mark for each correct word)

Unit 20.2

LL When **monomers** containing C=C double **bonds** are polymerised, no other molecule apart from the polymer is **formed**. We call this type of polymerisation **addition** polymerisation. An addition reaction is a reaction where **two** or more molecules **combine** and **no** other molecule is formed. [7]
(1 mark for each correct word)

1.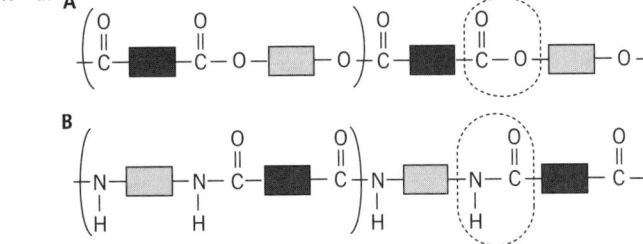

(1 mark for chain of 6 C atoms, 1 mark for rest of structure, 1 mark for continuation bonds on end carbon atoms) [3]

2.

$$\left[\begin{array}{cc} CH_3 & CH_3 \\ | & | \\ -C & -C- \\ | & | \\ H & H \end{array} \right]_n$$

(1 mark for structure, 1 mark for brackets and *n*, 1 mark for continuation bonds on end carbon atoms) [3]

3. Polymer A double bond [1] rest of structure correct [1]
Polymer B double bond [1] rest of structure correct [1]

Polymer **A**   Polymer **B**

$$\begin{array}{cc} C_6H_5 & H \\ | & | \\ C=C \\ | & | \\ H & H \end{array} \qquad \begin{array}{cc} H & CN \\ | & | \\ C=C \\ | & | \\ H & H \end{array}$$

Unit 20.3

LL In condensation polymerisation, molecules with different **functional** groups react together. In addition to the polymer a **small** molecule such as **water** or hydrogen **chloride** is **eliminated**. [5]
(1 mark for each correct word)

1. **a.** A

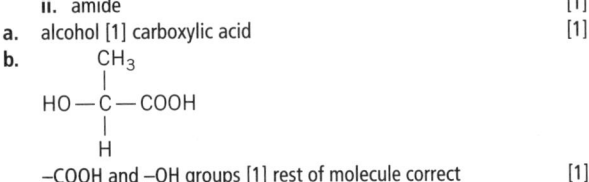

(1 mark for each) [2]

   b. i. ester [1]
   ii. amide [1]
2. **a.** alcohol [1] carboxylic acid [1]
   b.

$$HO-\overset{\displaystyle CH_3}{\underset{\displaystyle H}{\overset{|}{\underset{|}{C}}}}-COOH$$

   —COOH and —OH groups [1] rest of molecule correct [1]

Unit 20.4

LL Proteins are **condensation** polymers. They have the same **amide** linkage as nylon-6,6 but instead of being formed from two **different** monomers, proteins are made from about **twenty** different **amino** acids. There is no regular **repeat** unit in proteins because the amino acid residues are not in a regular **order**. [7]

1. **a.** Correct repeat unit e.g. first bracket between the first N and CHR from the left, second bracket between the next N and CHR to the right [1]
   b. amide / peptide [1]

2.

$$H_2N-\overset{\displaystyle H}{\underset{\displaystyle H}{\overset{|}{\underset{|}{C}}}}-C\overset{\displaystyle\diagup O}{\diagdown OH}$$

—COOH and —NH₂ groups [1] rest of structure correct [1]

3. Breakdown of a substance [1] using water (or acid or alkali as catalysts) [1]

Unit 21.1

1. **a.** Hydrocarbons are **compounds** containing **only** hydrogen and **carbon** atoms. (2 marks if all correct, 1 mark if 1 or 2 correct) [2]
   b. only [1]
2. **a.** A compound is a **substance** which contains **two** or more **different** atoms **bonded** together (2 marks if all correct, 1 mark if 1 or 2 correct) [2]
   b. different [1] bonded (or joined) [1]
3. **a.** Relative atomic mass is the **average** mass of the **isotopes** of an **element** compared to one-**twelfth** of the **mass** of an atom of ^{12}C.
   (3 marks if all correct, 2 marks if 3 or 4 correct, 1 mark if 1 or 2 correct) [3]
   b. Relative atomic mass is the **average mass** of the **isotopes** of an element compared to **one-twelfth** of the mass of an atom of ^{12}**C**.
   (3 marks if all correct, 2 marks if 3 or 4 correct, 1 mark if 1 or 2 correct) [3]
4. A mole is the **amount** of substance that contains 6.02×10^{23} specfied **particles** (atoms, ions, **molecules** or electrons) This number of particles is called the Avogadro **constant**.
(2 marks if all correct, 1 marks if 2 or 3 correct) [2]

Unit 21.2

1. minimum not crossed out [1]
2. type / kind / sort [1]
3. atoms not crossed out [1] element not crossed out [1]
4. decomposition / breakdown [1] electricity / electric current [1]
5. losing not crossed out [1]
6. ions not crossed out [1]
7. increases [1] volume not crossed out [1] frequency not crossed out [1]
8. ions not crossed out [1] move [1]
9. atom not crossed out [1] atom not crossed out [1]
10. components not crossed out [1] physical not crossed out [1]
11. double not crossed out [1] orange not crossed out [1] colourless not crossed out [1]

Unit 21.3

1. A with **5**, B with **3**, C with **6**, D with **1**, E with **4**, F with **2** [3]
   (3 marks if all correct, 2 marks if 4 or 5 correct, 1 mark if 2 or 3 correct)
2. A with **7**, B with **6**, C with **1**, D with **5**, E with **2**, F with **3**, G with **4** [4]
   (4 marks if all correct, 3 marks if 5 or 6 correct, 2 marks if 3 or 4 correct, 1 mark if 1 or 2 correct)
3. A with **6**, B with **4**, C with **5**, D with **1**, E with **3**, F with **2** [3]
   (3 marks if all correct, 2 marks if 4 or 5 correct, 1 mark if 2 or 3 correct)

Unit 21.4

1. A with **5**; B with **7**; C with **1**; D with **3**; E with **2**; F with **4**; G with **6** [4]
   (4 marks if all correct, 3 marks if 5 or 6 correct, 2 marks if 3 or 4 correct, 1 mark if 1 or 2 correct)
2. Diamond has a high melting point …. because it takes a lot of energy …. to break the strong …. covalent bonds which exist …. between all the carbon atoms.
(2 marks if all correct, 1 mark if one pair incorrect) [2]
3. The higher the temperature, …. the faster the particles move …. because they have more energy and collide …. with a greater frequency. The number of particles having energy …. equal to or greater than the activation energy …. also increases, so there is …. more chance of collisions between …. reactant particles being successful.
(1 mark for first sentence correct, 2 marks for second sentence correct – but 1 mark if one pair incorrect in second sentence) [3]

Unit 21.6

1. a. reactants [1] products [1] enthalpy [1] change [1]
   b. Describe [1] separate [1] dyes [1]
   c. Describe [1] observations [1] acid [1] zinc [1]
   d. Describe [1] explain [1] energy [1] movement [1] particles [1]
2. a. The word 'all' has not been considered [1] Metals such as potassium are soft and have relatively low melting points [1]
   b. No observation has been made [1] Observations are what you see (or hear or feel) / hydrogen gas is naming a substance (not an observation) [1]
   c. The name of a strong acid has been given [1] instead of the definition (an acid which is completely dissociated in solution) [1]
   d. This is not a feature of the molecule [1] The presence of a C=C double bond is a feature of the molecule / a test has been given instead [1]

Unit 22.2

1. a. Suggest [1]
   b. Describe [1] Name [1]
   c. Deduce [1]
   d. Describe [1] Explain [1]
   e. Draw [1] Determine [1]

Unit 22.4

1. Any three suitable (1 mark each) e.g. uses of particular chemicals / properties of selected elements / chemical tests e.g. water, ions, unsaturation / types of chemical reaction e.g. neutralisation, condensation, addition, photochemical [3]
2. Any four suitable with result (1 mark each) e.g.
   Metal hydroxides $\rightarrow$ salt + water
   Metal oxide $\rightarrow$ salt + water
   Carbonate $\rightarrow$ salt + carbon dioxide + water
   Methyl orange $\rightarrow$ turns red / pink
   Ammonia $\rightarrow$ ammonium salt formed
   Taste $\rightarrow$ sour (although this is not recommended!)
   What makes them acid? $\rightarrow$ hydrogen ions [4]
3. (1 mark for each reaction or property and 1 mark for each correct result e.g.) [16]
   alkane + chlorine in presence of light $\rightarrow$ chloroalkane + HCl
   alkane + (excess) oxygen / air $\rightarrow$ carbon dioxide + water
   alkane + limited oxygen $\rightarrow$ carbon monoxide + water
   alkane (heat with Al_2O_3) $\rightarrow$ mixture of alkanes and alkenes
   alkene + bromine water $\rightarrow$ bromine water decolourised
   alkene + hydrogen (in presence of Ni catalyst) $\rightarrow$ Alkane
   alkene + steam (in presence of catalyst) $\rightarrow$ alcohol
   alkene + (excess) oxygen / air $\rightarrow$ carbon dioxide + water

Unit 22.6

1. A low melting point [1]
   B/ C/ D High melting point [1] Conduct electricity when molten [1]
   Soluble in water [1]
   E or I metal [1] covalent giant structure [1]
   For the metal (E or I) (1 mark each for any 3 of:) Conduct electricity or conduct heat / ductile / malleable / lustrous (shiny) IGNORE: high melting point / sonorous [3]
   For the covalent giant structure (E or I) (1 mark each for any 3 of:) High melting point / generally do not conduct electricity / insoluble in water [3]
2. A increase in rate [1] B concentration [1] C increase rate [1]
   D The particles collide with greater frequency [1]
   E Increases rate [1] F surface area [1] G increase rate [1]
   H Increased number of particles exposed for collisions / increased collision frequency [1]

Unit 23.1

1. a. 2 Na and 1 O [1] b. 3 Mg and 2 N [1] c. 1 P and 3 Cl [1]
   d. 2 Al and 3 O [1] e. 2 H, 1 S and 4 O [1]
2. a. 4 H and 2 S [1] b. 15 Mg and 10 N ([1]
   c. 6Al and 9 S [1] d. 6 Fe and 8 O [1] e. 6 Li, 3 C and 9 O [1]
3. a. $(2 \times 27) + (3 \times 16) = 102$ [1]
   b. $(2 \times 23) + 12 + (3 \times 16) = 106$ [1]
   c. $207 + 32 + (4 \times 16) = 303$ [1]

Unit 23.2

1. a. 1 Sn, 2 S, 8 O [1]
   b. 2 N, 8 H, 1 S, 4 O [1]
   c. 1 Ni, 2 Cl, 8 O [1]
   d. 1 Ba, 2 I, 6 O [1]
2. a. $117 + (2 \times 32) + (8 \times 16) = 309$ [1]
   b. $(2 \times 14) + (8 \times 1) + 32 + (4 \times 16) = 132$ [1]
   c. $59 + (2 \times 35.5) + (8 \times 16) = 258$ [1]
   d. $137 + (2 \times 127) + (6 \times 16) = 487$ [1]
3. $59 + (2 \times 35.5) + 6 \times (2 + 16) = 238$ [1]

Unit 23.3

1. a. actual yield $= \dfrac{\% \text{ yield}}{100} \times$ theoretical yield [1]
   b. theoretical yield $= \dfrac{\text{actual yield}}{\% \text{ yield}} \times 100$ [1]
2. a. moles = concentration (in mol / dm³) × volume (in dm³) [1]
   b. volume (in dm³) $= \dfrac{\text{moles}}{\text{concentration (in mol / dm}^3)}$ [1]
3. mass = density × volume [1]

Unit 23.4

1. a. 1×10^6 [1]   b. 70 000 [1]
   c. 3.3×10^3 [1]   d. 3200 [1]
2. a. 1×10^{-6} [1] b. 0.005 [1]
   c. 3.5×10^{-3} [1] d. 0.25 [1]
3. a. 1.4 [1]   b. 3.6×10^{-5} [1]
4. a. 1.14×10^{-5} [1] b. 2.67×10^2 [1]

Unit 23.5

1. a. 87.7 % [1]
   b. mass pure substance = 3.45 – 0.12 kg = 3.33 kg [1]
   % purity = 96.5 % [1]
2. a. i. 6 [1]
      ii. 96 cm² [1]
   b. $5 \times 5 \times 5 = 125$ cm³ [1]

Unit 23.6

1. a. i. 4.36 [1]   ii. 0.0873 [1]
      iii. 137 [1]   iv. 0.00550 [1]
   b. i. 440 [1]   ii. 3.4 [1]
      iii. 57 [1]   iv. 0.0055 [1]
2. mol pentane 0.0926388 [1]   rounded 0.09 [1]
   × 5   0.4631944 [1]   rounded 0.45 [1]
   × 24   11.1 [1]   10.8 [1]

Unit 23.7

1

Axes labelled correctly [1] full use of graph paper [1]
positive values correct [1] negative values correct [1]

Answers

Unit 23.8

1. a. Line not continued to 0-0 point [1] line not a continuous curve [1]
      line has more points above it than below it [1]
   b. Points not clear [1] no units on *y* or *x* axis [1] full grid not used / lot of grid is space [1]
   c. Anomalous point included in the line [1] straight lines between points / not a smooth curve [1]

2. a. 26 cm³ [1] b. 44–44.2 cm³ [1]

Unit 23.9

1.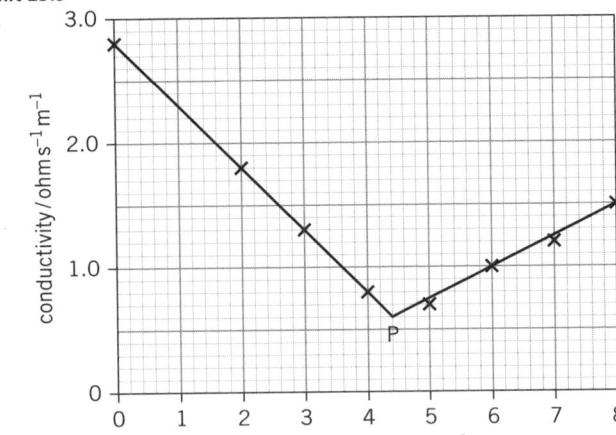

   a. Axes correctly labelled [1]
      Points all correct (1 mark if one point incorrect or missing) [2]
   b. Lines correct (two straight intersecting lines) [1]
      Lines intersect between 4 and 5 cm³ and P labelled [1]
   c. P is 4.4 cm³ [1]

2.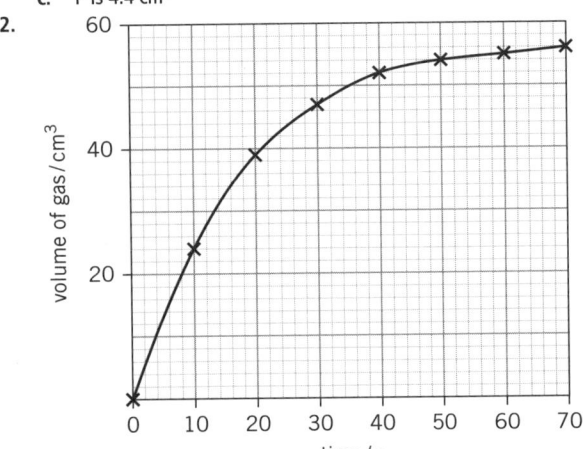

   Axes correctly labelled and full grid used [1]
   Points all correct (1 mark if one point incorrect or missing) [2]
   Smooth curve between the points [1]

Unit 23.10

1.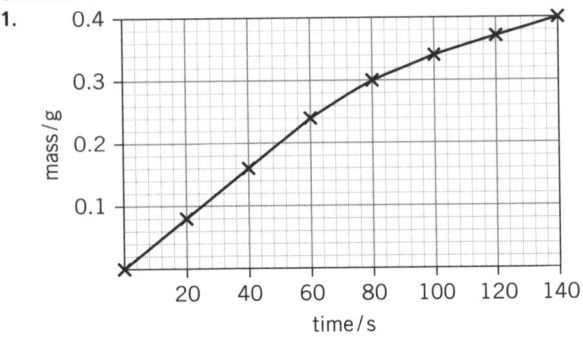

Unit 24 (right column)

   a. Axes correctly labelled and full grid used [1]
      Points all correct (1 mark if one point incorrect or missing) [2]
      Smooth curve between the points [1]
   b. 0.16/40 [1] 4×10^{-3} g/ s [1]
   c. The line starts to curve / the rate is not constant [1]

Unit 24.1

1. Goggles / safety glasses / eye protection [1]
2. Carry out in a fume cupboard / hood [1] Use gloves / labcoat / eye protection [1]
3. A Explosive [1] B Corrosive [1] C Toxic / Poisonous [1] D Flammable [1]
   E Harmful [1]
4. a. Flammable [1] b. Corrosive [1] c. Toxic / Poisonous [1]
5. Funnel has no tap [1] gas would escape from this [1]
   Tube of water [1] sulfur dioxide is soluble in water / reacts with water [1]
   gas jar wrong way up [1] hydrogen chloride is heavier than air [1]

Unit 24.2

1. Apparatus to be used [1] quantities or concentrations to be used [1]
   conditions to be used e.g. heat [1] measurements or observations to be made [1] what to control and what to vary (idea of fair test) [1]
2. a. temperature [1]
   b. volume of carbon dioxide [1]
   c. (1 mark each for any two of:) time / mass of calcium carbonate / surface area of calcium carbonate / concentration of acid [2]
3. a. mass of fuel burnt [1]
   b. temperature rise [1]
   c. (1 mark each for any two of:) volume of distance of burner from can / same copper can / same volume of water in can / same height of flame or same wick used in burner [3]

Unit 24.3

1. Repeating the results several times until consistent results are obtained [1]
   Using equipment which measures accurately e.g. burette instead of measuring cylinder [1]

2.

volume	temperature
0.0	0
4.0	2
8.4	6
12.2	10

(For each column: 2 marks if all correct, 1 mark if 2 or 3 correct) [2]

3. S = 34 cm³ [1]   T = 15 cm³ [1]

Unit 24.4

1.

time / s	0	10	20	30	40	50	60	70	80	80
mass change / g (first set)	0.0	0.5	0.89	1.24	1.44	1.56	1.68	1.79	1.82	1.82
mass change / g (first set)	0.0	0.72	1.04	1.24	1.44	1.52	1.62	1.74	1.78	1.78
average / g	0.0	0.61	0.965	1.24	1.44	1.54	1.65	1.765	1.80	1.80

(1 mark for time, masses and average in column 1, 1 mark for correct units, 1 mark for correct values, 1 mark for correct averages) [4]

2.

temperature / °C	time / s	$\dfrac{1}{\text{time}}$ / s⁻¹

(1 mark for temperature, time and rate, 1 mark for correct units) [2]

Unit 24.5

1. a. Independent: time [1] Dependent: conductivity [1]
   b. Independent: concentration of acid [1] Dependent: time [1]

2. a.

(1 mark for each line correct) [4]
Axes of graph correctly labelled [1]

b. The one at 3 minutes for the concentration of 3.2 mol / dm³ [1]
It does not fall on the same line as the other points / it is lower than expected [1]

Unit 24.6

1. a. i. A, B and E [1]
ii. Doubling the concentration doubles the rate / rate is directly proportional to concentration
Increasing the concentration increases the rate (alone) [2]
b. i. B, C and D [1]
ii. Doubling the concentration doubles the rate / rate is directly proportional to concentration
Increasing the concentration increases the rate (alone) [2]
c. With only 2 different concentrations, a consistent pattern cannot be found [1]
Idea that it needs at least 3 points to show proportionality / show doubling concentration doubles the rate [1]

2. Deduce the gradient of each line [1] from the slope of relative conductivity divided by time [1] plot rate against concentration (to give straight line through origin) [1]

Unit 24.7

1. a. As pressure increases, volume decreases [1]
The rate of decrease, decreases with increasing volume ALLOW volume is inversely proportional to pressure / the graph is a concave downwards curve [1]
b. Volume of gas is proportional to concentration of acid [2]
(If 2 marks not scored: Volume of gas increases linearly with concentration of acid = 1 mark)
c. Rate doubles for every 10 °C rise [2]
(If 2 marks not scored: Rate gets faster as temperature increases = 1 mark)

Unit 24.8

1. a. thermometer [1] beaker [1] top pan balance and weighing boat [1]
water bath with temperature control ALLOW Bunsen burner and tripod and gauze [1] stirring rod [1]
b. Independent variable: temperature of the water [1]
Dependent variable: mass of solid added [1]
c. (1 mark each for any two of:) volume of the water / rate at which heat is applied e.g. from water bath or Bunsen / rate of stirring [2]

2. Place 50 cm³ / 100 cm³ water in beaker [1]
Heat water to fixed temperature [1]
Add small weighed amounts of solid potassium chloride to the water and stir [1]
Keep adding small weighed amounts until no more dissolves [1]
Repeat at different temperatures [1]

3. a. (1 mark each for any two of:) If using Bunsen, the water will cool while the substance is being added / If using water bath, temperature control depends on sensitivity of thermostat / Heat may be given out or absorbed when substance dissolves in water. [2]
b. (1 mark each for any two of:) Use a temperature-controlled water bath OR more sensitive temperature control / Use larger volume of water so that inaccuracies due to adding small amounts of solid are reduced / Use insulated container so heat losses reduced / Add solid to the water until saturated solution formed, then allow to cool and take temperature when crystals first appear. [2]
c. Temperature change [1], because thermometer reading precise to 0.5 °C at best and temperature change is only a few degrees, whereas mass can be measured to a precision of two or three decimal places [1]

Unit 24.9

1. a. burette / measuring cylinder / gas syringe [1] flask [1] stopper/ bung and connecting tubing [1] stopclock [1] top pan balance and weighing boat [1]
b. Independent variable: either: (i) time (to collect given volume of gas) or (ii) volume of gas (collected in a given time) [1]
Dependent variable: If (i) volume of gas / If (ii) time [1]
c. mass / moles of each metal oxide [1] concentration of hydrogen peroxide [1] volume of hydrogen peroxide / temperature / size of metal oxide particles [1]

2. Small amount of oxide / stated values e.g. 0.1 to 1 g [1] suitable volume of hydrogen peroxide e.g. 1–50 cm³ [1] Add to flask and start stopclock [1] Measure volume of gas collected in fixed time / Measure time (to collect fixed volume of gas) [1] Repeat under same conditions with another metal oxide [1]

3. a. Difficult to measure volume accurately when liquid level / syringe plunger moving [1]
Difficult to read stopclock and volume at same time [1]
b. Get someone else to read the stopclock or volume [1]
Use more sensitive balance (reading to 3 or 4 decimal places) [1]

Exam-style questions

Unit 26

1. a. 2,8,5 [1]
b. 16 [1]
c. Does not conduct electricity / does not conduct heat [1]
Low melting point / low boiling point [1]
d. P_4 (or 4P) + $5O_2 \rightarrow 2P_2O_5$
(1 mark for correct formulae, 1 mark for balance) [2]
e. PO_4^{3-} [1]
f. Nitrate [1]
g. Fertilisers needed for plant growth / for plant proteins [1]
Sources of nitrogen / phosphorus / potassium in soil used up by growing plants [1]
h.

1 pair of electrons shared between each of the 3 H atoms and the central P atom [1]
lone pair on the P atom [1]

2. a. double C=C bond [1]
b. bromine water / bromine ALLOW acidified potassium manganate(VII) [1]
decolourised [1]
c. i. reduction is gain of electrons [1]
ii. high temperature [1] high pressure [1] catalyst [1]

iii.

(2 marks for full structure, 1 mark if OH drawn instead of O – H) [2]

   d. i. Grind up the onion leaves in a solvent / water / alcohol [1]
         filter [1]
      ii. chromatography [1]

3. a. $CsCl$ / Cs^+Cl^- [1]
   b. giant structure / ionic structure [1]
      All the bonds are strong / strong electrostatic forces between
      all ions [1]
      (mention of atoms / intermolecular forces = maximum 1
      for question)
   c. The ions are free to move (from place to place) [1]
   d. i. Conduct electricity [1]
         unreactive / inert [1]
      ii. anode: $2Cl^- \rightarrow Cl_2 + 2e^-$ [2]
      (1 mark for formulae, 1 mark for balance)
      cathode: $Cs^+ + e^- \rightarrow Cs$ [1]
   e. mol Cs = 5.32 / 133 = 0.04 mol [1]
      actual yield of CsCl = 6.4/ 168.5 = 0.038 [1]
      % yield = 0.038 / 0.04 = 95% [1]
      (or calculation based on masses)

4. a. 6.5 min [1]
   b. i. 16 cm³ [1]
      ii. 26/2 = 13 cm³/ min [1]
   c. Initial gradient steeper [1] Ends up at the same volume of gas [1]
   d. Rate increases (0 mark) but if missing or incorrect maximum 1 mark
      for the question. Increasing concentration increases the number of
      particles per unit volume / particles closer together [1]
      Frequency of collisions increases / number of collisions
      per second increases [1]
   e. Faster because greater surface area of powder [1]
      More particles of magnesium exposed to hydrochloric acid [1]

5. a. decomposition [1]
      endothermic [1]
   b. Bubble through limewater [1] limewater turns milky / cloudy [1]
   c. i.

      Axes correctly labelled [1] points plotted correctly [1] curve of
      best fit drawn [1]
      ii. mass of CO_2 from graph = 3.0 g [1]
         moles CO_2 = 3/44 = 0.068 mol [1]
         volume = 0.068 × 24 = 1.64 dm³ [1]

6. a. i. Arrow under the flask pointing upwards [1]
      ii. **A** flask [1] **B** gas jar [1]
      iii. To dry the ammonia / to remove water [1]

      iv. Put damp red litmus beneath gas jar [1]
         Full when (litmus) turns blue [1]
   b. $(NH_4)_2SO_4 + 2NaOH \rightarrow 2NH_3 + Na_2SO_4 + 2H_2O$
      (1 mark for correct formulae, 1 mark for balance) [2]
   c.

      (1 mark for bonding pairs of electrons, 1 mark for the lone pairs on
      each nitrogen atom) [2]

7. a. Circle around the O – H group [1]
   b. i. (1 mark each for any two of:) carbon, hydrogen and oxygen [2]
      ii. alcohols [1]
   c. carbon dioxide [1] water [1]
   d. filtration [1]
   e. i. ester [1]
      ii. no continuous carbon chain [1] idea of the COO groups being
         formed by condensation reactions / COO group formed from
         condensation of alcohol and carboxylic acid [1]
   f. purple [1] to colourless [1]

8. a. i. helium <u>and</u> neon [1]
      ii. ALLOW values between 0.08 and 0.1 [1]
      iii. gas [1] −118 is above the boiling point [1]
      iv. Increases down the Group [1]
   b. i. brown [1] ALLOW grey / black [1]
      ii. (1 mark each for any two of:) iodide is being converted to
         iodine / oxidation number of iodine increases / oxidation
         number of Xe decreases [2]
      iii. mol XeF_4 = 8.28 / 207 = 0.04 mol [1]
         0.04 mol Xe [1]
         0.04 × 24 = 0.96 dm³ Xe [1]

9. a. i. car exhausts / high temperature furnaces / lightning [1]
      ii. acid rain / kills trees / acidifies lakes / erodes limestone /
         corrodes metal structures etc [1]
      iii. Proximity: far apart [1] Motion: fast / random [1]
   b. i. Colour gets lighter [1] position of equilibrium moves to
         the left [1]
         in direction of fewer gas molecules / fewer moles in
         the equation [1]
      ii. NO_2 46 [1] N_2O_4 = 92 [1]
      iii. Entirely NO_2 at 140 °C / more NO_2 at higher temperature [1]
         The higher the temperature the more the equilibrium goes to
         the right [1]
         For an endothermic reaction the position of equilibrium
         moves to the right with increase in temperature / increase in
         temperature favours the endothermic reaction [1]
   c. $2NO_2 \rightarrow 2NO + O_2$
      (1 mark for correct formulae, 1 mark for balance) [2]

10. a. Any suitable indicator e.g. methyl orange / litmus /
      thymolphthalein [1]
   b. sodium sulfate [1]
   c. i. (12.5 / 1000) × 0.2 = 2.5×10^{-3} mol [1]
      ii. 5.0×10^{-3} mol [1]
      iii. 5.0×10^{-3} × 1000/25 = 0/20 mol / dm³ [1]
   d. $H^+ + OH^- \rightarrow H_2O$ [1]
   e.

      (2 marks if all correct, 1 mark if ester group shown as COO) [2]

The Periodic Table of Elements

Group

Key

atomic number
atomic symbol
name
relative atomic mass

I	II	III	IV	V	VI	VII	VIII
							2 **He** helium 4

| 1 **H** hydrogen 1 | | | | | | | |

| 3 **Li** lithium 7 | 4 **Be** beryllium 9 | 5 **B** boron 11 | 6 **C** carbon 12 | 7 **N** nitrogen 14 | 8 **O** oxygen 16 | 9 **F** fluorine 19 | 10 **Ne** neon 20 |
| 11 **Na** sodium 23 | 12 **Mg** magnesium 24 | 13 **Al** aluminium 27 | 14 **Si** silicon 28 | 15 **P** phosphorus 31 | 16 **S** sulfur 32 | 17 **Cl** chlorine 35.5 | 18 **Ar** argon 40 |

19 **K** potassium 39	20 **Ca** calcium 40	21 **Sc** scandium 45	22 **Ti** titanium 48	23 **V** vanadium 51	24 **Cr** chromium 52	25 **Mn** manganese 55	26 **Fe** iron 56	27 **Co** cobalt 59	28 **Ni** nickel 59	29 **Cu** copper 64	30 **Zn** zinc 65	31 **Ga** gallium 70	32 **Ge** germanium 73	33 **As** arsenic 75	34 **Se** selenium 79	35 **Br** bromine 80	36 **Kr** krypton 84
37 **Rb** rubidium 85	38 **Sr** strontium 88	39 **Y** yttrium 89	40 **Zr** zirconium 91	41 **Nb** niobium 93	42 **Mo** molybdenum 96	43 **Tc** technetium –	44 **Ru** ruthenium 101	45 **Rh** rhodium 103	46 **Pd** palladium 106	47 **Ag** silver 108	48 **Cd** cadmium 112	49 **In** indium 115	50 **Sn** tin 119	51 **Sb** antimony 122	52 **Te** tellurium 128	53 **I** iodine 127	54 **Xe** xenon 131
55 **Cs** caesium 133	56 **Ba** barium 137	57–71 lanthanoids	72 **Hf** hafnium 178	73 **Ta** tantalum 181	74 **W** tungsten 184	75 **Re** rhenium 186	76 **Os** osmium 190	77 **Ir** iridium 192	78 **Pt** platinum 195	79 **Au** gold 197	80 **Hg** mercury 201	81 **Tl** thallium 204	82 **Pb** lead 207	83 **Bi** bismuth 209	84 **Po** polonium –	85 **At** astatine –	86 **Rn** radon –
87 **Fr** francium –	88 **Ra** radium –	89–103 actinoids	104 **Rf** rutherfordium –	105 **Db** dubnium –	106 **Sg** seaborgium –	107 **Bh** bohrium –	108 **Hs** hassium –	109 **Mt** meitnerium –	110 **Ds** darmstadtium –	111 **Rg** roentgenium –	112 **Cn** copernicium –		114 **Fl** flerovium –		116 **Lv** livermorium –		

lanthanoids

| 57 **La** lanthanum 139 | 58 **Ce** cerium 140 | 59 **Pr** praseodymium 141 | 60 **Nd** neodymium 144 | 61 **Pm** promethium – | 62 **Sm** samarium 150 | 63 **Eu** europium 152 | 64 **Gd** gadolinium 157 | 65 **Tb** terbium 159 | 66 **Dy** dysprosium 163 | 67 **Ho** holmium 165 | 68 **Er** erbium 167 | 69 **Tm** thulium 169 | 70 **Yb** ytterbium 173 | 71 **Lu** lutetium 175 |

actinoids

| 89 **Ac** actinium – | 90 **Th** thorium 232 | 91 **Pa** protactinium 231 | 92 **U** uranium 238 | 93 **Np** neptunium – | 94 **Pu** plutonium – | 95 **Am** americium – | 96 **Cm** curium – | 97 **Bk** berkelium – | 98 **Cf** californium – | 99 **Es** einsteinium – | 100 **Fm** fermium – | 101 **Md** mendelevium – | 102 **No** nobelium – | 103 **Lr** lawrencium – |

Relative atomic masses (A_r) for calculations

Element	Symbol	A_r
aluminium	Al	27
bromine	Br	80
calcium	Ca	40
carbon	C	12
chlorine	Cl	35.5
copper	Cu	64
fluorine	F	19
helium	He	4
hydrogen	H	1
iodine	I	127
iron	Fe	56
lead	Pb	207
lithium	Li	7
magnesium	Mg	24
manganese	Mn	55
neon	Ne	20
nitrogen	N	14
oxygen	O	16
phosphorus	P	31
potassium	K	39
silver	Ag	108
sodium	Na	23
sulfur	S	32
zinc	Zn	65